Echolocation in
Whales and Dolphins

Echolocation in
Whales and Dolphins

P. E. Purves
Formerly of British Museum (Natural History), London, England

G. E. Pilleri
Hirnanatomisches Institut, Ostermundigen (Bern), Switzerland

1983

ACADEMIC PRESS

A Subsidiary of Harcourt Brace Jovanovich, Publishers

London   New York
Paris   San Diego   San Francisco   São Paulo
Sydney   Tokyo   Toronto

ACADEMIC PRESS INC. (LONDON) LTD.
24/28 Oval Road
London NW1

United States Edition published by
ACADEMIC PRESS INC.
111 Fifth Avenue
New York, New York 10003

British Library Cataloguing in Publication Data
Purves, P. E.
    Echolocation in whales and dolphins
    1. Echolocation (Physiology)
    I. Title   II. Pilleri, G. E.
    599.01'825   Q469

    ISBN 0-12-567960-2
    LCCCN 82-72328

Typeset by Bath Typesetting Ltd., Bath
and printed in Great Britain by
Galliard (Printers) Ltd., Great Yarmouth

Preface

During 1966 W. E. Evans wrote in his paper on *Vocalization Among Marine Mammals**—"It is safe to say that the complexity of cetacean vocalizations is exceeded only by the fervour of the research efforts of cetologists and behaviourists to explain their function".

Since many of the large and small whales and dolphins have been reduced by Man to near extinction there has been an increasing awareness by the general public of the urgency to conserve the Cetacea as a whole. As a result a great number of popular and semi-scientific books have been published assigning almost human-like attributes to these interesting animals. Notable among these is the idea that they possess a complicated communication system or language similar to that of Man.

It has even been postulated that a knowledge of "delphinese" may one day help us to communicate with some alien creature from outer space! It may be stated, however, that Man is probably the only mammal on Earth which possesses such a language to communicate with members of his own species about events which occurred in the past or may occur in the future. Recently, the celebrated anthropologist Dr Richard E. Leakey postulated, amidst much controversy that *Homo erectus* who lived $1\frac{1}{2}$ to 2 million years ago was the true ancestor of modern Man. Leakey also postulated that *Homo erectus* was capable of articulate speech (Leakey and Lewin, 1977). For anatomical reasons, which will be explained later, the idea was probably correct and this fact alone could be responsible for Man's pre-eminence over all other living creatures. At the same time it was his greatest tragedy since he alone could predict his eventual death, and for this reason developed a spiritually based culture.

*In "Marine Bio-Acoustics" (W. N. Tavolga, ed.). Pergamon Press, Oxford–London, 1967.

Cetaceans like the Microchiroptera (bats) live predominantly in a world of echoes, so that their sonar emission must necessarily be monitored by echoes which will differ from the emitted sounds in quality and carry information about the objects producing the echoes. The voice is for this purpose isolated acoustically from the ears as in the bats but in quite a different manner. Cetaceans emit a variety of sounds which are within the human auditory range and these are probably heard by the phonating animal though somewhat imperfectly. These sounds will be referred to only briefly in this book. No neurophysiological experiments have been carried out by the authors for a variety of reasons. The bony structures in the vicinity of the essential organ of hearing are so robust as to require general anaesthesia lasting up to 20 hours. Surgery requires the complete ablation of so many vital parts that to date all neurophysiological experiments have ended in the death of the animals concerned. The animals are required to be kept afloat in saline solutions so that any electrophysiological "cochlea microphonic" results must be viewed with extreme circumspection. The size and complexity of the brain indicates that small cetaceans occupy a position high up in the hierarchical system of mammals and surgical experiments on living animals seem to be too high a price to pay in order to prove or disprove a minor physiological point. One eminent Professor of neuroanatomy after a visit to our laboratories wrote, "Would it not be nice if, somewhere, some "mini-dolphin" existed?—a kind that one could conveniently keep and breed in a laboratory and use for electrophysiological and neuroanatomical experiments, enabling one to obtain the type of data on brain organization of this sophisticated group of mammals—much as one used the (small, cheap and manipulable) marmosets in order to gain insight into the primate brain (though there of course one can use big monkeys and even apes as well)".

One of the purposes of this book is to demonstrate that the results of simple experiments on models, fresh specimens, and acoustical measurements with living animals under natural conditions are entirely consistent with the anatomical facts and could be arrived at by deductive reasoning without vivisection which is always undesirable in such large and intelligent animals. For the sake of accuracy anatomical terms have been used but these may be obtained from any standard text-book of Human Anatomy.

September, 1982   P. E. Purves
                                                        G. E. Pilleri

Acknowledgements

Grateful acknowledgments are extended to various individuals who have been of assistance in the accumulation of the published data and may be named in order of their participation in our investigations: the late Dr F. C. Fraser, CBE, FRS, British Museum (Natural History), for his collaboration during the very early stage of the research; Dr W. L. van Utrecht, University of Amsterdam; Dr J. C. Lilly, Communications Research Institute, Miami, Florida; the Swiss National Fund for the Promotion of Scientific Research in Berne; Dr M. Gihr, Dr C. Kraus, Dr K. Zbinden, Brain Anatomy Institute, University of Berne; Mrs S. Jennings, Geneva; Dr S. Andersen, University of Odense, Denmark; Mr Nathar Khan, Gizri Village, Karachi. We wish to thank Mrs T. Tüscher, Berne, for her valuable assistance in the typing of the manuscript.

Contents

IV. Dolphin Sounds (*Sousa plumbea*)

V. Sonar Fields (*Inia geoffrensis*)

VI. Properties of the Sonar System of Cetaceans with Pterygoschisis: *Delphinapterus leucas, Neophocaena phocaenoides, Phocoena phocoena*

VII. Phonation in the blind Indus Dolphin *Platanista indi*

VIII. The Ear of Cetaceans

Introduction

At an "Interdisciplinary Symposium on Animal Sonar Systems" held during 1979 in Jersey, Channel Islands, under the auspices of and financed by the North Atlantic Treaty Organization, one speaker introduced his lecture thus—"Previous to the work of our group which was published in 1970, all theories of hearing in Cetacea were based on dissections and experiments with dead specimens. Such experiments continue to be published in the literature to this day, and just as the earlier studies of dead material did, they only serve to confound the many students and investigators of Cetacean hearing, especially those who have little formal training in the science of physiological acoustics".

Perhaps the term "electrophysiological acoustics" would have been better, as this group were concerned only with the processing of mechanical vibrations at the very end of the acoustical pathways, i.e. where such vibrations are turned into electrical activity suitable for reception by the brain. None of the electrophysiological experiments carried out to date have contributed anything to our knowledge of the *sources* of emission or reception of sound by cetaceans.

One eminent American zoologist has propounded a number of theories concerning sound production and reception in cetaceans based on purely intuitive or subjective observation. It is clear that these hypotheses have been arrived at with only a rudimentary application of anatomy and none of acoustics. Nevertheless a great number of elaborate and costly experiments have been mounted in the United States in order to prove that this zoologist was correct in his assumptions.

One typical example of how a preconceived notion can influence even the study of anatomy was afforded by another group, who made an elaborate

study of cross sections of the head and larynx of seven dolphins which
had died of natural causes in the wild, in oceanariums or in their own
laboratories. In their very next, relatively short paper on electromyography
and pressure events in the nasopharyngeal system of dolphins during
phonation, the same group immediately after the Introduction stated—
"There were no pressure changes in the trachea. This rules out the larynx
as a primary sound source". Had their anatomical studies been discerning
enough, they would have found that no intratracheal pressure could be
expected during sonar emission in cetaceans.

The present authors have found cetacean tissues to be particularly
suitable for sound conduction experiments since they seem to be remarkably free from pathogenic micro-organisms of any kind (especially specimens from the polar regions) until such organisms are introduced into
the body by Man.

Provided the tissues are reasonably fresh and the experiments carried
out under sterile conditions, they are found to transmit sound practically
without attenuation. If a 100 kHz sound wave is passed through a small
bowl of fresh whale oil there is no appreciable attenuation. If the oil is
blown into a froth and then allowed to stand until there remain no bubbles
visible to the naked eye the sound transmission is found to have been
reduced by 80%.

It is because so many impractical theories of sound emission and
reception in cetaceans have been widely publicized and apparently
accepted, that we have found it necessary to begin this book with a
reiteration of some of the most elementary laws relating to sound.

I. Transmission of Sound in Air and Water

I. Reflection and Refraction

When a wave of sound meets the boundary surface between two different media it is partially reflected, and a wave travels in the negative direction through the incident medium with the same velocity as it approached the boundary.

The geometrical laws of reflection of sound waves are the same as apply to light-waves, the angles of incidence and reflection being equal and in the same plane. In many cases, however, the length of the sound waves is comparable with the linear dimensions of the reflecting objects, when the geometrical laws cease to apply and the phenomena must be regarded essentially as diffraction. A hypothetical case to be considered is one in which a source is emitting spherical waves within a closed cavity, the dimensions of which are comparable with the wavelength of the emitted sound. As the walls are inelastic and consist of parabolic, spherical and plane surfaces of an extended medium the geometry of reflection and diffraction is exceedingly complex. We can, however, apply the normal laws of energy reflection. These are obtained without assumptions regarding wavelength and phase and are applicable to waves of any type whatsoever.

The geometrical laws of reflection and refraction follow directly from the facts that the velocity in each medium is independent of the direction of the wave front and that traces of all wave fronts on the plane of separation have equal velocities. Consequently if $\theta_1, \theta_r \theta_2$ are the angles of incidence, reflection and refraction respectively (Fig. 1) measured with respect to the

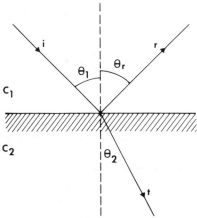

Fig. 1   Diagram showing reflection and refraction of sound waves at an interface between two media of different densities.

"normal" at a line drawn perpendicular to the reflecting surface:

(1)
$$\frac{\sin \theta_1}{c_1} = \frac{\sin \theta r}{c_1} = \frac{\sin \theta_2}{c_2}, \text{ or } \theta_1 = \theta r$$

which is the law of reflection and:

(2)
$$\frac{\sin \theta_1}{\sin \theta_2} = \frac{c_1}{c_2}$$

the law of refraction where c_1 and c_2 are the velocities of sound in the first and second media respectively.

The amount of *energy* reflected at normal incidence $\theta_1 = \theta_2 = 0$ is given by the equation:

(3)
$$\frac{R_1 - R_2}{R_1 + R_2} = r$$

where R_1 and R_2 are the radiation resistances or "acoustic impedances" of the first and second media respectively.

(4)
$$R = c\rho$$

where c is the velocity of sound through the medium and ρ the density of the medium.

(5) The amount of energy transmitted is given by:

$$\frac{2R_1}{R_1 + R_2} = t_{12}$$

If the sound wave now be transmitted from medium (2) to medium (1), we have:

(6)
$$\frac{R_2 - R_1}{R_1 + R_2} = r$$

II. Phase Changes on Reflection

Referring to equation 3 it is important to observe that when $R_2 = R_1$ there is no reflected wave and transmission is at maximum. When $R_2 > R_1$ the acoustic impedance in the first medium being less than the second (e.g. air to water) there is no change of phase. When $R_2 < R_1$ (e.g. water to air) there is a reversal of phase. The phase of the transmitted wave remains the same in both cases.

When the sound travels from a medium of low velocity c_1 to a medium of higher velocity (i e. $C_2 > C_1$) we see from equation (1) that there will be a *critical angle* of incidence θ_1 when θ_2 becomes 90° that is,

$$\text{when } \sin \theta_1 = \frac{C_1}{C_2}$$

beyond which there is total reflection. On account of the greater velocity of sound in water, total reflection may occur when the waves are incident from air on water, but not from water to air.

To illustrate the application of the above equations, take the case of transmission from air to water. At normal incidence equations 3 and 6 are applicable. For air $R_1 = \rho_1 c_1 = 40$ and for water $R_2 = \rho_2 c_2 = 1 \cdot 5 \times 10^5$ whence the ratio of reflected and incident amplitudes is 0·9994 indicating almost total reflection.

From water to air the reflection coefficient is again almost unity. In practically all cases of sound transmission from air to a solid or liquid medium (or vice versa) i.e. *whenever the acoustic impedances are widely different there is almost complete reflection.* The difficulty of transmitting sounds from water (e.g. the noise of a ship's propeller) to the air-filled ear-cavity of an observer will be apparent.

The problem of incorporating air cavities in underwater sound transducers was discovered during the construction of early hydrophones. This was partially overcome by the use of energized carbon granules as in telephones. Modern high-powered hydrophones make use of magneto-

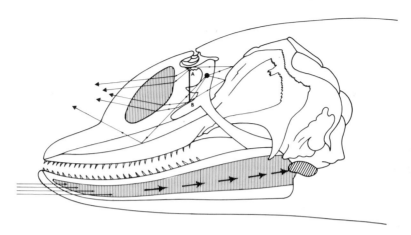

Fig. 2 Diagram after Norris showing the focussing effect of sound waves through an "acoustic lens" from a source in the upper nares of a dolphin. The figure also shows his hypothesis of sound reception through the lower jaw.

striction or in the case of low-powered high-frequency senders and receivers, piezoelectric materials which do not require an air cavity.

We may now apply the foregoing equations to Fig. 2 after Norris in which the "acoustic lens" hypothesis of sound emission in dolphins is illustrated. This hypothesis assumes the sound waves to be brought to a focus in front of the forehead or "melon" of the dolphin. The sound generator is seen in a closed air-cavity, in this case the right nostril the walls of which for the sake of simplicity are assumed to have the same density as that of sea water. Equations 1 and 3 are applicable where $\theta_1 = \theta_R$ and $R_1 = 40$ and $R_2 = 1 \cdot 5 \times 10^5$ and in this case only a minute fraction of emitted energy would penetrate the surrounding structures, the rest being almost totally internally reflected. If any of the pathways of sound happened to be equal to or an exact multiple in length with that of the emitted sounds, resonances would occur within the cavity and the system would go on vibrating indefinitely until the energy became dissipated in the form of heat, see Fig. 3.

In Fig. 4 the sound generator has been moved so that it is in acoustic contact with the posterior wall of the nostril. Sound radiated forward would be almost totally internally reflected as in Fig. 3. Some of the energy would be radiated posteriorly and penetrate the soft tissues without reflection. At the surface of the skull equations 1, 2, and 5 would apply; some of the energy would be reflected in the relationship $\theta_1 = \theta_R$ and some

transmitted and refracted in the relationship

$$\frac{\sin \theta_1}{\sin \theta_2} = \frac{C_1}{C_2}$$

where C_1, C_2 are the velocities of sound in the animal tissues and the bone

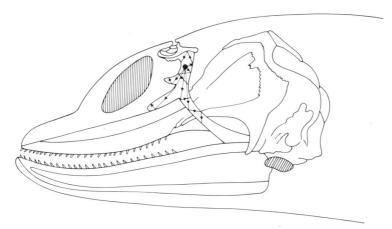

Fig. 3 Diagram showing total internal reflection of sound from a source in the same position as in Fig. 2. The reflection is due to the widely different acoustic impedances of the air in the nares and that of the animal tissues.

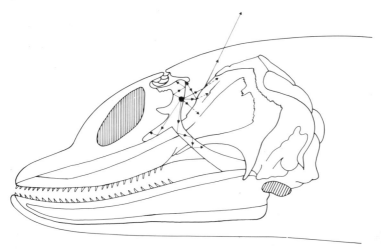

Fig. 4 The sound source has been moved so as to be in acoustic contact with the posterior wall of the nares. Some sound is transmitted through the soft tissue to the bone of the skull where some is transmitted and some reflected. Sound passing forward is totally internally reflected.

respectively. The amount of energy transmitted to the bone would be relatively small as it depends on the ratio of the acoustic impedances R_1, R_2 that of animal tissues and petrous bone respectively which in this case equal 1.5×10^5 and 4.1×10^5, so that a maximum of 34% of the energy is transmitted. This would be rather smaller where the angle of incidence is very large with respect to the normal.

In Fig. 5 the sound generator has been moved so that it is in acoustic contact with the anterior wall of the nostril. In this position the energy

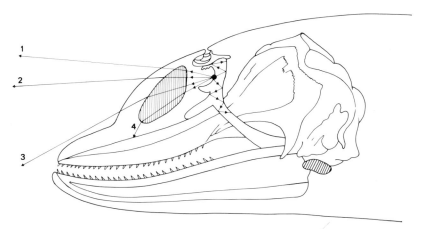

Fig. 5 The sound source has been moved so as to be in acoustic contact with the anterior wall of the nares. The sound waves are transmitted through the soft tissues to the "acoustic lens", the refractive index of the lens (glycerides of isovaleric acid) is not large enough to focus the sound waves to a point as suggested in Fig. 2. Sound passing backwards is totally internally reflected.

radiated backwards can be ignored but that radiated forward requires some detailed examination. In Fig. 2 all the energy passing forward through the "acoustic lens" is seen as being reflected from two air-filled cavities, the vestibular sac A, and the supramaxillary sac B, just above the rostrum of the skull.

In reality there would be no air in the vestibular sac since there is no means of preventing the air from escaping through the blowhole through hydrostatic pressure when the head is submerged. According to the hypothesis of a narial origin of sonar pulses air is "metered" out from the supramaxillary sac to the sound producing organ during a burst of sonar. In this event the sac would change shape and consequently so would the angles of reflection. In this diagram no refraction is seen at the anterior

surface of the lens but as the velocity of sound is purported to be lower in the substance of the lens than outside there would a substantial downward refraction at this surface.

In Fig. 5 four lines of sound radiation are depicted passing through the lens with no intervening air spaces. By producing an enlarged copy of this diagram it is possible to measure the angles of incidence of the four waves at the back and front faces of the lens respectively and using equation No 2 in which:

$$\sin\frac{\theta_1}{\theta_2} = \frac{V_1}{V_2}$$

V_1 for tissue surrounding lens $= 1\cdot5 \times 10^5$ cm/s, V^2 for substance of lens, glycerides of isovaleric acid $= 1\cdot37 \times 10^5$ cm/s, the angles of refraction can be obtained. These are as follows:

	Angle of incidence back of lens	Angle of refraction inside lens	Angle of incidence front of lens	Angle of refraction outside lens
Wave No. 1	$12°$	$11°$	$51°$	$59°$
Wave No. 2	$0°$	$0°$	$43°$	$49°$
Wave No. 3	$18°$	$16°$	$50°$	$57°$
Wave No. 4	$60°$	$52°$	$13°$	$14°$

It will be seen that none of these waves are focussed to a point in front of the head as postulated—indeed it would be highly undesirable if they did so since refraction is independent of frequency over the whole range of sounds known to be emitted by cetaceans so that all waves would be focussed to the same point whatever their frequency. The sound source is too small and too near the lens for the latter to have any significant focussing effect on the sound waves. The authors of Fig. 2 have attempted to surmount this problem by depicting two sound reflecting surfaces at A and B so that the sound source becomes larger with respect to the lens but the angles of reflection and refraction remain erroneous. It should also be pointed out that there are a number of species of echolocating cetaceans in which vestibular and premaxillary air-sacs are vestigeal or absent.

We must therefore look for a sound source which does not require an "acoustic lens" and a transmission path to the sea water which does not incur a serious acoustic mismatch at any point. It will be shown later that the melon has important hydrodynamic functions and may also act as an organ of dynamic orientation.

The present authors regard the larynx as the primary source of sound— as it is in every other mammal including the echolocating Microchiroptera—the bats. The sound transmission from this organ is, however, strongly influenced by an elaborate system of foam-filled sinuses associated with the middle ear. The main function of these air sinuses is to adjust the air pressure in the middle ear to that of the ambient sea water since this can obviously not be done by swallowing. Since diverticula of these sinuses envelope the essential organ of hearing, the cochlea and the muscles conducting sound from the larynx, they have a considerable effect in isolating the ear from the voice—a requirement which is essential in echolocating mammals.

III. Velocity in Air–Water Mixtures

The velocity of sound through a mixture of non-resonant air bubbles in water is given by:

$$(7) \qquad c = \sqrt{\frac{E}{\rho}} = \sqrt{\frac{E_1 E_2}{\{xE_2 + (1-x)E_1\}\{x\rho_1 + (1-x)\rho_2\}}}$$

In this case we have:

for air— $\rho_1 = 0.0012$; $E_1 = 1.2 \times 10^6$
for water— $\rho_2 = 1.0$; $E_2 = 2.25 \times 10^{10}$

Employing these values in Equation 7 for the mean velocity and substituting various values of x (the fraction of air by volume) we obtain the curve shown in Fig. 6. Here it is seen that a proportion of one part air in 10 000 parts water lowers the velocity of sound by about 40%. The curve also reveals that the velocity reaches a minimum value when the volume of air is equal to that of water. This minimum velocity is only about 1/15 (22 m/s) of the velocity of sound in air alone. Under such conditions the mixture may be regarded as foam, the water merely serving to load the air bubbles (the springs), and consequently the velocity of transmission. It will be seen from the curve, as is otherwise obvious, that the velocity becomes equal to the velocity in water or air respectively as x is equal to zero or unity.

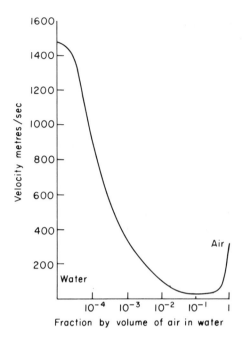

Fig. 6 Graph showing the reduction of the velocity of sound waves through a mixture of air and water.

The damping being due to the large volume variation in the liquid surrounding each bubble, the former may be calculated in the case of a spherical bubble surrounded by a sphere of viscous (but incompressible) fluid subjected to a simple harmonic variation of radius. The efficiency of the bubbles in damping vibrations increases rapidly both as their diameter and distance apart diminishes. This may be verified experimentally. Consequently the efficiency of the foam in the accessory air sinuses of the cetacean ear increases with depth, notwithstanding the diminution in volume of the sinuses.

IV. Reflection from Air–Water Mixtures

The acoustic impedance of a mixture of air and water is given by the equation:

(8)
$$\rho c = \sqrt{\frac{E_1 E_2 \{x\rho_1 + (1 - x)\rho_2\}}{x E_2 + (1 - x)E_1}}$$

where E_1 and E_2 represent the elasticity of air and water respectively and x is the proportion of air by volume. Using the value ρc in the equation:

(9)
$$r^2 = \left(\frac{\rho c - \rho_2 c_2}{\rho c + \rho_2 c_2}\right)^2$$

where $\rho_2 c_2$ refers to clear water, the percentage energy reflected r^2 is plotted as a function of x in Fig. 7.

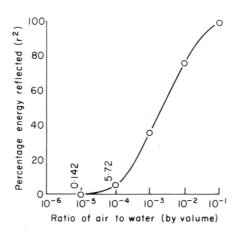

Fig. 7 Graph showing the percentage of sound energy reflected at the interface from a mixture of air and water.

The energy is partially reflected and the rest almost totally absorbed by the foam. As has been pointed out the cochlea in cetaceans is completely enveloped by foam-filled cavities so that there can be very little sound transmission by bone conduction. In all the electrophysiological experiments on cetaceans involving cochlea potentials these foam-filled cavities have been destroyed so that the cochlea has been subject to vibration from extraneous sources.

V. General Hypothesis of Phonation and Hearing in Cetaceans

Figure 8 is a schematic representation of the sonar mechanism propounded by the present authors. Vibrations of the cartilages of the larynx EPG, CUC, CAR, are transferred to the palatopharyngeal muscles MPP(S),

MPP, and directed forward to the insertions of these muscles on the palatine bones at the posterior face of the rostrum of the skull where the faciculi of the muscle fibrils enter the longitudinal spaces between the trabeculae of the bone. Thence the vibrations are radiated in the manner shown to the blubber and sea water. The sound intensities at various points on the skull can be measured experimentally and are always at a maximum along the mid-line (see pp. 43–49). It will be seen that the area of transmission to the skull constitutes a *double* source so that outside the limits of the skull the two sound fields overlap and interfere with one another to produce directional beams of sound (see pp. 65–66). This principle is widely used in the construction of radio direction-finding equipment such as radar.

In most of the pelagic cetaceans the two sets of muscles MPP are also enveloped by foam-filled air sinuses (Fig. 8: PTS). In the more neritic species there is a gap between these sinuses on the ventral surfaces of the muscles referred to in this book as *pterygoschisis*. In these species the sound is found to radiate downwards from the throat as well as forward, and has an important bearing on the food finding habits of these animals.

One of the most surprising features of the foam in the air sinuses is its behaviour under high pressure. The sample investigated was placed in a pressure-tight optical cell and observed by transmitted light through a vertically mounted, low-power microscope with micrometer eyepiece. Pressure was applied by pumping B.P. liquid paraffin into the cell (in direct contact with the foam) and was measured on a Bourdon gauge. As pressure was increased the foam bulk volume decreased and was replaced by a system of spherical bubbles dispersed in the liquid. These bubbles were a few microns in diameter and separated by distances of comparable magnitude. The system was stable at higher pressures and the bubbles had not disappeared after 20 min at 100 atm. On release of pressure to 1 atm the foam structure reappeared. Using egg albumen, the foam structure was again replaced by air bubbles dispersed in a continuous liquid phase, which persisted at 100 atm. The bubbles were of less regular size and shape than in the cetacean foam. Using detergent, the foam structure collapsed and no bubbles were visible at higher pressures. From the results of the experiments just described it may be deduced that, even at the greatest depths to which cetaceans normally dive, air bubbles would persist and there would be a sound damping and reflecting system surrounding the essential organ of hearing.

The rate of change of bulk volume of the foam would be important because of the variation in pressure in rapid changes of depth. Acoustic

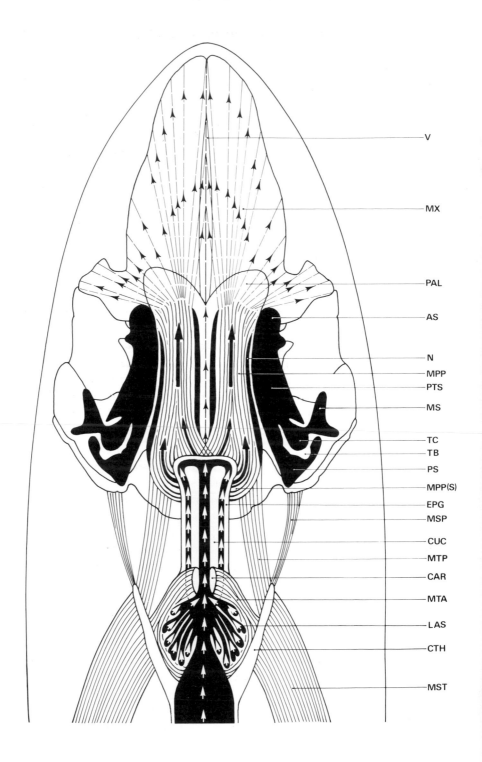

V

MX

PAL

AS

N
MPP
PTS

MS

TC
TB
PS
MPP(S)
EPG
MSP

CUC

MTP

CAR

MTA

LAS

CTH

MST

efficiency must ultimately depend on the maintenance of fairly constant conditions of sound reflection and absorption round the essential organ of hearing.

The air-sacs themselves are so extensive that their contraction would cause disruption of the adjacent musculature were it not for the intervention of some space filling mechanism. This is provided by the injection of blood into a fibrovenous plexus which envelopes the air sinuses. Its origin and structure will be described when dealing with the anatomy of the middle ear.

Figure 9 is a schematic representation of the mode of hearing in cetaceans, and a typical land mammal. Owing to the large difference in acoustic impedance between air and animal tissues (Fig. 9, A), over 99% of the energy of sound waves striking the side of the head of a land mammal is reflected. Only that which enters the ear orifice can be converted to a form acceptable to the sound analyser, the cochlea, which is normally fused to and forms part of the skull.

This means that a sound shadow is created on the side away from the source and the noise is less intense on that side. Because of this, the animal knows whether the sound source is right or left and can move the ears so that the maximum difference is obtained at the two ears. This, together with the difference in time of arrival and quality of the sound waves received by each ear accurately determines the direction of the source.

The acoustic impedance of water and animal tissues (Fig. 9, B) is very nearly the same and in these circumstances most of the sound energy which strikes the side of the head and body of a cetacean passes straight through without reflection.

Fig. 8 Diagram showing the coupling of sound vibration generated by the epiglottic spout to the palatopharyngeal muscles and skull. Black areas denote air spaces. White areas denote cartilage or bone. White arrows show direction of air streams. Black arrows show propagation of sound waves.

AS = air-sac, CAR = arytenoid cartilage, CTH = thyroid cartilage, CUC = cuneiform cartilage, EPG = epiglottis, LAS = laryngeal air-sacs, MPP = palatopharyngeus muscle, MPP(S) = palatopharyngeal sphincter, MS = middle sinus, MSP = salpingopharyngeal muscle, MST = sternothyroid muscle, MTA = thyroarytenoid muscle, MTP = thyropharyngeus muscle, MX = maxilla, N = naris, PAL = palatine bone, PS = posterior sinus, PTS = pterygoid sinus, TB = tympanic bulla, RC = tympanic cavity, V = vomer.

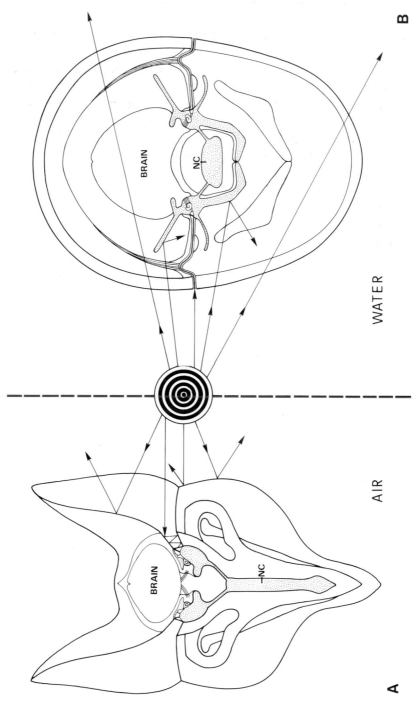

WATER

AIR

B

A

Fig. 9 (A–B)   Schematic drawing showing sound transmission and reflection in a typical and land mammal (A) and an odontocete cetacean (B). NC = nasal cavity.

In the absence of special arrangements, the sound intensity at both ears would be equal and the animal would be unable to determine the direction of the source. This occurs when the human head is immersed in water. For this reason the cochlea in the cetacean is isolated acoustically from the cranium by foam-filled air spaces which are continuous with those on the side of the skull. These spaces are seen in cross-section in the diagram. By reflecting or absorbing all the sound energy except that which enters the ear orifice, the air spaces cast acoustic shadows which produce intensity and quality differences at the ears. Since the velocity of sound in water is four times greater than that in air, time of arrival differences at the two ears probably play very little part in cetacean directional hearing.

It will be seen from the foregoing preliminary account that (to use American terminology) the present authors *reject* the hypothesis of a narial origin of sonar signals in cetaceans and the theory that the "melon" acts as an "acoustic lens" to produce focussed, directional beams of sound.

II. The Anatomy of Phonation in Cetaceans

During the acquisition of the upright stance in the course of the evolution of Man the whole of the base of the respiratory tract including the lungs has lost the support of the rib-cage and sternum and has become increasingly under the influence of the force of gravity. Vertical support is provided only by the diaphram which is relatively more muscular in Man than in most other mammals.

As a consequence, all the muscles of the pharynx and indeed of the entire neck have become subject to stretching forces and therefore considerably elongated dorso-ventrally. The larynx too has been removed further from the mouth cavity than in any other mammal. This statement sounds like a Lamarckian concept, but since this process has conferred considerable survival advantages on Man, it is probably the result of natural selection.

The net result has been the production of a great number of resonant cavities above the level of the larynx which are employed in the formation of speech and singing. The so-called vocal chords cannot alone produce speech, their chief function being simply to alter the pitch of the voice. During phonation the vocal chords are closely approximated to form an acute triangular slit aperture (Fig. 10) through which during phonation a large number of "vortices" of air are expelled. Some or all of these vortices can be made to resonate in the pharynx and mouth by careful control of the constituent muscles, notably those of the tongue and lips, to produce the speaking or singing voice.

A great number of mammals do not possess vocal chords and therefore cannot alter the pitch of the voice. Griffin (1958) was of the opinion that the pulsing of the bat's echolocating signal was carried out at the level of the epiglottis but that the frequency modulation within the pulse was

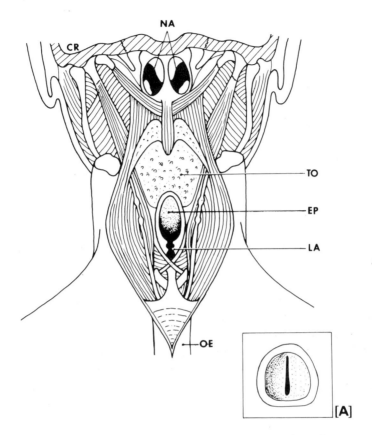

Fig. 10 Schematic drawing of the human pharynx during phonation as seen from behind showing the ventral separation of the larynx from the nasal passages and the great length of the pharyngeal muscles. The shape of the aperture between the vocal chords during singing as seen from below (A).
CR = cranium, EP = epiglottis, LA = larynx, NA = nasal apertures, OE = oesophagus, TO = tongue.

performed by true vocal chords within the larynx. We believe that the pulsing of the cetacean echolocating signal is also carried out at the level of the epiglottis. There is no frequency modulation of the type found in bats.

During the evolution of the Cetacea quite the reverse process has occurred in the modification of the nasopharynx. Living as they do in an entirely aquatic environment, the whole body is less subject to gravitational forces than in any other mammal. The seven vertebrae of the neck are greatly shortened antero-posteriorly, sometimes wafer-thin in small cetaceans and generally fused together into a solid mass.

The muscles of the pharynx are also extremely short and fused together to form the palatopharyngeal sphincter. However, by resection and careful dissection of the sphincter all the components forming the nasopharynx in

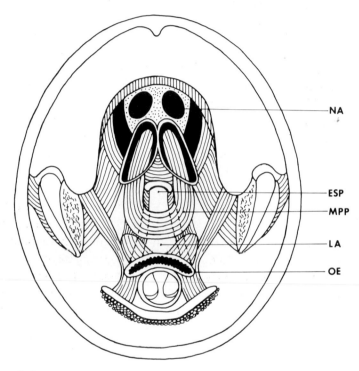

Fig. 11 Schematic drawing of the nasopharynx of a cetacean during phonation as seen from behind showing the epiglottic spout and the absence of a pharyngeal cavity.

ESP = epiglottic spout, LA = larynx, MPP = palatopharyngeal sphincter, NA = nasal apertures, OE = oesophagus.

Man can be revealed (Boenninghaus, 1902; Purves and Pilleri, 1978). The muscles forming the side walls of the mouth are rudimentary or absent. Thus there are no resonant cavities associated with the mouth and pharynx (Fig. 11). During deglutation in the Cetacea the glottis is withdrawn from the palatopharyngeal sphincter when large food objects are being swallowed (Pilleri, 1979).

Resonant cavities of any kind are undesirable in underwater phonation because of the large acoustic mismatch between sound vibrations in air and water.

I. Structure of the Larynx

Since Tyson (1680) first described the larynx of the common porpoise, *Phocoena phocoena*, a great number of anatomists have interested themselves in the cetacean vocal organ, and at the present time the larynges of very few genera remain to be described (Murie, 1870, 1873; Pouchet and Beauregard, 1889; D'Arcy Thompson, 1892; Rawitz, 1900; Benham, 1901; Boenninghaus, 1902; Schulte, 1916; Kernan and Schulte, 1918; Hosokawa, 1950; Kleinenberg and Yablokov, 1958; Brown, 1962; Sukhovskaya and Yablokov, 1979). More recently a comprehensive study of the larynx has been made by Green *et al.* (1979) in which many of the main structures have been divided into smaller units with a new anatomical nomenclature. In their very next paper in the same volume the authors have "ruled out the larynx as being the primary sound source". It would seem therefore that all this detailed work was a waste of time! It is inconceivable to us that such complexity of structure could be used for mere ventilation of the lungs.

Throughout this book the anatomical nomenclature has been taken from Grays' Anatomy. The general form (Fig. 12) is typically mammalian and consists of a skeletal framework of cartilages held together by ligaments, muscles and mucous membranes. It is possible to identify all the main structures by the names applied to their homologues in human anatomy, and in describing the modifications that have occurred the larynx of Man will be taken as a reference.

The most striking difference in the form of the cetacean larynx from that of all other mammals lies in the structure of the glottis, which is in the shape of an elongated spout. The distal end of the spout, the axis of which is orientated antero-dorsally to that of the rest of the larynx, is inserted into the posterior nares (Fig. 13). Much confusion has arisen in the interpretation of the structures inside the larynx, owing to an assumption that the dorsal inclination of the glottis (GLO) is a secondary development, correlated with the dorsal situation of the blowhole (BLO). In fact, dorsal orientation of the glottis is a feature common to most mammals, and it is the horizontal orientation of the trachea (TRA), cricoid and thyroid (Fig. 14: CTH) cartilages, which is the secondary development, due to

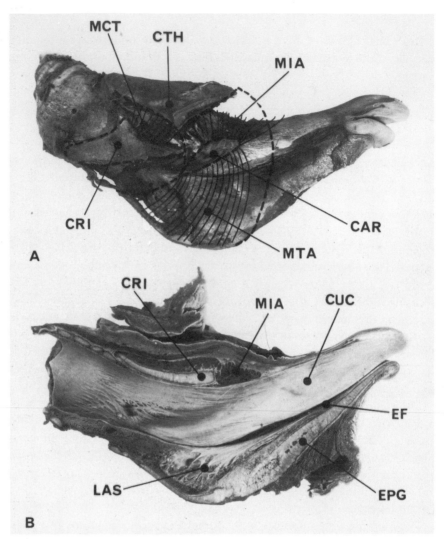

Fig. 12 (A–B) (A) The larynx of a common dolphin, *Delphinus delphis*, showing muscles and cartilages. The greater part of the thyroid cartilage has been removed to show underlying muscles which are active during phonation. (B) Bisected larynx to show the great length of the epiglottic and cuneiform cartilages which form the epiglottic spout and the relationship between the thyroarytenoid muscle and the laryngeal air-sacs.

CAR = arytenoid cartilage, CRI = cricoid cartilage, CTH = thyroid cartilage, CUC = cuneiform cartilage, EF = epiglottic fold, EPG = epiglottis, LAS = laryngeal air-sacs, MCT = cricothyroid muscle, MIA = interarytenoid muscle, MTA = thyroarytenoid muscles.

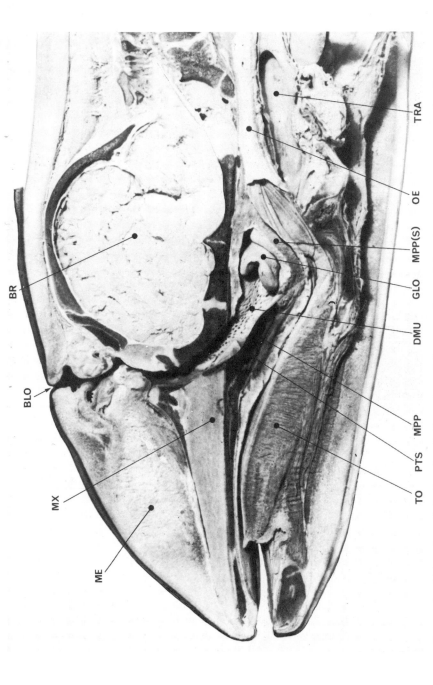

Fig. 13 Parasagittal bisection of the head and dissection of the posterior narial region of a common porpoise *Phocoena phocoena* showing the relationship between the glottis and the palatopharyngeal sphincter. Part of the connection between the palatopharyngeal muscle and bones of rostrum is also shown. BLO = blowhole, BR = brain, DMU = mucous ducts, GLO = glottis, ME = melon, MPP = palatopharyngeal muscle, MPP(S) = palatopharyngeal sphincter, MX = maxillary bone, OE = oesophagus, PTS = pterygoid sinus, TO = tongue, TRA = trachea.

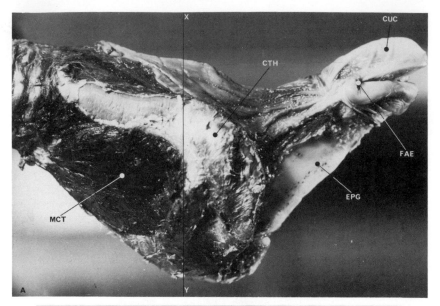

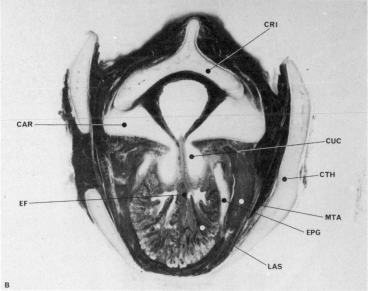

Fig. 14 (A–B)   (A) Larynx of the bottle-nosed dolphin, *Tursiops truncatus*, showing the thyroid cartilage intact and the aryepiglottic fold which is the assumed point of emission of the sonar pulses. (B) Cross-section through the larynx of the bottle-nosed dolphin through the line *xy* (in (A)) to show the relationship between the thyroarytenoid muscles and the laryngeal air-sacs.
CAR = arytenoid cartilage, CRI = cricoid cartilage, CTH = thyroid cartilage, CUC = cuneiform cartilage, EF = epiglottic fold, EPG = epiglottis, FAE = aryepiglottic fold, LAS = laryngeal air-sacs, MCT = cricothyroid muscle, MTA = thyroarytenoid muscle.

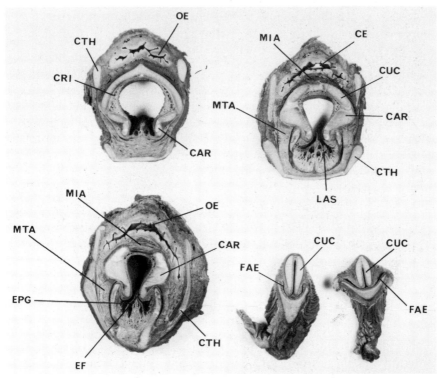

Fig. 15   A series of transverse sections through the larynx of *Phocoena phocoena*
from the level of the cricoid cartilage to the tip of the epiglottic spout showing
the spout divided into three narrow channels at its apex. The two lateral channels
are assumed to produce synchronous sonar pulses and the central channel,
high-pitched whistles.

CAR = arytenoid cartilage, CRI = cricoid cartilage, CTH = thyroid cartilage,
CUC = cuneiform cartilage, EF = epiglottic fold, EPG = epiglottis, FAE =
aryepiglottic fold, LAS = laryngeal air-sac, MIA = interarytenoid muscle,
MTA = thyroarytenoid muscle, OE = oesophagus.

shortening of the neck and respiratory tract. The sharp horizontal inclination of the respiratory tract at the level of the thyroid cartilage has
caused the arytenoids, (Figs 14, 15 : CAR) to be more closely intruded into
the thyroid angle, so that the thyroarytenoid muscles (Figs 14, 15 : MTA)
are relatively short compared with those of other mammals of comparable
size. For the same reason the cricoid (CRI) is partially telescoped over the
arytenoids and the ligamentous attachment is greatly augmented, so that
the arytenoids are not free to rotate on the cricoid as in terrestrial mammals.

In the angle of the thyroid cartilage and between the two thryoarytenoid muscles a remarkable system of bilaterally symmetrical air chambers (Figs 14, 15: LAS), lined by mucous membrane and mucoid tissue, completely fills the vocal part of the rima glottidis, isolating the thyroarytenoid muscles from the latter. Thus there are no vocal folds nor vocal chords of the type found in terrestrial mammals. These pouches end in blind diverticula posteriorly but continue anteriorly as far as the base of the epiglottis (EPG) where they break into a symmetrical arrangement of high, longitudinal folds (Fig. 15: EF) formed from the mucous lining of the epiglottis. The base of the epiglottis (Figs 14, 15: EPG) is also curiously modified, taking the form of a deep trough, the lateral walls of which reach posteriorly well beyond the anterior borders of the arytenoid (CAR) and cuneiform (CUC) cartilages and, enclosing the system of epiglottic folds, described above. Because of this arrangement, the aryepiglottic folds (FAE) are deep, extremely narrow and supported both mesially and laterally by stiff cartilage. More distally, in the upper portion of the epiglottic spout, the longitudinal folds become gradually narrower and less elevated, until only the medial fold remains. The greater part of the epiglottic trough in this region is filled by thick, hyaline cartilage. In the Ziphiioid whales (Fig. 16) the medial fold is also replaced by cartilage. Anteriorly to the cricoid, the fused arytenoid and cuneiform cartilages which, with the interarytenoideus muscle, form the posterior boundaries of the epiglottic spout, become more and more approximated to each other and to the trough of the epiglottis, so that the anterior part of the rima glottidis tapers gradually until, at the distal end of the spout, it remains only as a minute aperture. Where all three cartilages touch, the mucous lining is perfectly smooth, so that when the glottis is constricted, no interval remains between the cartilages. If the epiglottic spout of any delphinid larynx is cut into transverse sections from the base upwards, the morphological arrangement of the epiglottic folds, described above, can be seen clearly. In the region of the air chambers, the medial fold is but slightly elevated, so that the interval between the anterior borders of the arytenoid cartilages is fairly wide. At the level of the base of the epiglottis, the medial fold is very high and completely fills the arytenoid interval. At this point it is also flanked by two less prominent, tapering folds (the vocal folds of Murie) and a number of shallower folds. As the medial fold is traced up the epiglottic spout it is found to follow the contours of the anterior borders of the cuneiform cartilages (CUC) exactly, filling the interval between them.

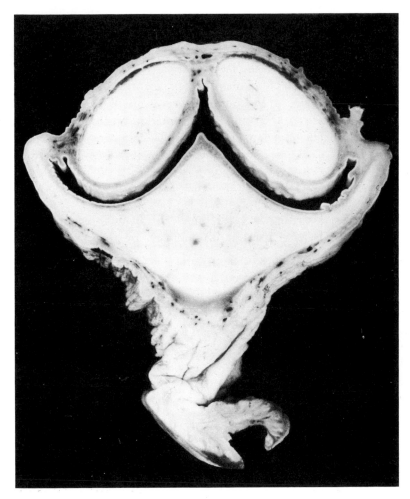

Fig. 16 Cross-section of larynx of *Mesoplodon bidens*.

By the whole arrangement described above, the upper part of the rima glottidis can, when the glottis is constricted, be divided into three narrow channels, one lying between the cueneiform cartilages and the other two bounded by the epiglottis, cuneiform cartilages and the aryepiglottic folds. At the very apex of the epiglottic spout the cuneiform cartilages taper abruptly and become embedded in two rounded bosses of yellow elastic tissue. The epiglottic cartilage widens laterally into a flange, which curls posteriorly and dorsally, partially embracing the cuneiforms, to which it is joined by the lateral aryepiglottic folds (Fig. 14: FAE).

Notwithstanding the extreme complexity of the air-sac system and its associated epiglottic folds and cartilages, the whole assembly is remarkable for its bilateral symmetry, so that when the epiglottis is constricted and air blown through the trachea, the larynx expands in a perfectly symmetrical manner. Owing to the thin, membranous nature of the folds and the posterior operculation of the pouches, a system of anteriorly directed valves operates to drive the contained air forward up the glottis when the larynx is constricted by manual pressure.

II. Structure of the Nasopharynx

As previously stated, the tip of the epiglottic spout is inserted into the posterior nares. At this point it is embraced by the palatopharyngeal

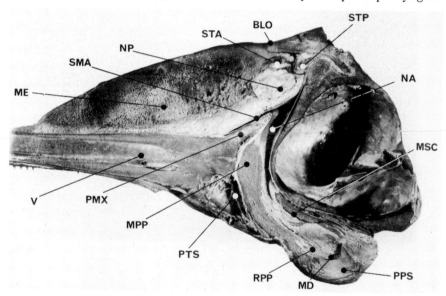

Fig. 17 Section through the epiglottic spout of the beaked whale, *Mesoplodon bidens*, showing the relationship of the massive palatopharyngeus muscle to the pterygoid sinus and rostrum. The relationship of the tubular air-sacs to the nasal plug is also shown.

BLO = blowhole,   MD = mucous   ducts,   ME = melon,   MPP = palato-pharyngeus muscle, MSC = superior constrictor muscle, NA = naris, PMX = premaxilla, PPS = palatopharyngeal sphincter, PTS = pterygoid sinus, RPP = palatopharyngeal recess, SMA = supramaxillary air-sac, STA = anterior tubular sac, STP = posterior tubular sac, V = vomer.

sphincter and the arcus palatinus. Owing to the sharp horizontal flexure of the respiratory tract (TRA) and the intranarial situation of the epiglottic spout, the pharynx and oesophagus (OE) are orientated postero-dorsally from the oral cavity, so that nearly the whole of the arcus palatinus surrounds the posterior nares, and lies within the mesial borders of the palatopharyngeal sphincter. In this position it is difficult to separate the two sets of muscles, and in dissection it is necessary to trace them from their points of origin. The arcus palatinus commences as two sets of muscles high in the posterior nares (Figs 13, 17, 18), the pars interna (MPP[I]) of the palatopharyngeus being by far the most massive in each

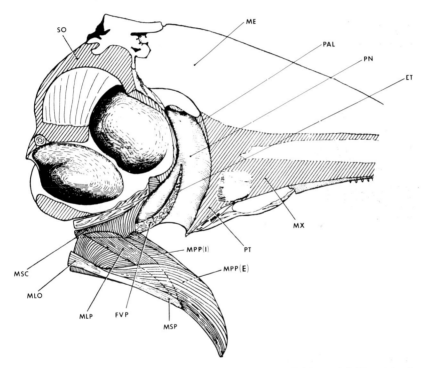

Fig. 18   Diagram of a dissection of the head of *Delphinus delphis* with the palatopharyngeal muscle mass reflected to show its constituent parts which are the equivalent of the muscles of the human pharynx shown in Fig. 10.
ET = eustachian tube, FVP = fibrovenous plexus, ME = melon, MLO = longitudinal oesophageal muscle, MLP = levator palati muscle, MPP(E) = external palatopharyngeus muscle, MPP(I) = internal palatopharyngeus muscle, MSC = superior constrictor muscle, MSP = salpingopharyngeus muscle, MX = maxilla, PAL = palatine bone, PN = posterior naris, PT = pterygoid.

set. They originate on the posterior aspect of the palatine bone, and the upper surface of the palatine aponeurosis. Passing postero-ventrally on each side of the vomer, they are in relation to the nares (PN) posteriorly and to the pterygoid bones and their air spaces (PTS), anteriorly, laterally and ventrally. Below the nares, the fibres of each muscle decussate both dorsally and ventrally and surround the epiglottic spout, forming a deep semi-circular recess (Fig. 17: RPP) into which are inserted the lips of the epiglottis and lateral aryepiglottic folds. The pars externa (MPP[E]), of the palatopharyngeus originate more ventrally on the palatine aponeurosis and descends in company with the pars interna. Below the nares they also decussate dorsally and ventrally and with the superior constrictor (MSC) form part of the dorsal and external rim of the palatopharyngeal sphincter. It is obvious that the pars interna of the palatopharyngeus referred to above corresponds to the pterygopharyngeus of Lawrence and Schevill (1965) and that the pars externa corresponds to their palatopharyngeus. If these names are to be assigned to the separate parts of the palatopharyngeus, we prefer to reverse the order and refer to the pars interna as the palato-pharyngeus since it clearly originates on the palatine bones. The pars externa on the other hand originates on the postero-dorsal aspect of the pneumatized pterygoid bones and is therefore more properly called the pterygopharyngeus. From the functional point of view this distinction is important, since the acoustic coupling of the palatopharyngeus with the rostrum is very good, whereas that of the pterygopharyngeus is poor because of its contiguity with the pterygoid air spaces. As Lawrence and Schevill point out, the pars interna forms the strong sphincter round the tip of the "arytenoepiglottic cartilages". The pars externa is continuous with the "pars thyropalatina" which inserts between the thyroid and the epiglottic cartilages. It follows, therefore, that the acoustic coupling between the tip of the epiglottic spout and the bones of the rostrum will be good whilst that between the thyroid cartilages and the rostrum will be poor.

The anterior portion of the outer sheath of the palatopharyngeal sphincter is largely non-muscular and consists of a strong fibrous band, which is continuous with the palatine aponeurosis along the whole posterior border of the pterygoid hamuli. It sweeps backward, lateral to the levator palati near its upper border. The mucous lining of the posterior nares, which covers the whole system of muscles, is marked by a regular arrange-ment of mucous ducts whose apertures are directed postero-ventrally towards the palatopharyngeal sphincter. It is important to note that beyond the level of the sphincter the two sets of intranarial muscles diverge in a symmetrical manner, and are almost wholly enveloped in air spaces connected with the middle ear.

The palatopharyngeal muscles are in relation to these air-sac systems on their ventral and lateral aspects, and to the nasal passages on their posterior and dorsal aspects, along nearly their entire lengths. It is only along the central axis dorsally and at the origins of the posterior surfaces of the palatine bones that there is no contiguous air space. To complete the general description of the nasopharynx, it must be stated that as with the larynx, this part of the respiratory tract is perfectly bilaterally symmetrical and in marked contrast with the upper narial region, which in the odontocete is noted for its asymmetry. Even in the Sperm-whale *Physeter catodon*, in which one of the upper nares is from five to seven times greater in diameter than the other, there is not the least trace of asymmetry in the posterior narial region.

III. Architecture of the Skull

Although it is appreciated that the minute architecture of the skull is mainly related to the manner of growth of the component bones, it is believed by the authors that it also has an important influence on the distribution of sound waves emanating from the larynx, so a brief description is required here.

If the ventral surface of the skull of any odontocete is examined closely, it will be seen that from the arcuate, anterior borders of the palatine bones (Fig. 19, A) the trabeculae spread fan-wise on the ventral surface of the rostrum and the orbital processes of the frontal bones (B). A great number of the trabeculae on each side converge on the central axis of the rostrum (C) and would ultimately meet, were it not for the intrusion of the intermaxillary suture and vomer. The trabeculae of the vomer itself are all very nearly parallel with the long axis of the rostrum. Examination of the disarticulated bones of a young specimen shows that the corrugations on the frontomaxillary suture (D) also diverge from the palatine region. In other regions of the skull, particularly those which are contiguous with air spaces, the matrix of the bones appears to be more compact and the trabeculae less uniformly orientated. It is interesting to note too, that the axes of the foramina, distributing nerves and blood vessels to the surface of the rostrum (A), all diverge from the palatine region in general conformity with the orientation of the trabeculae.

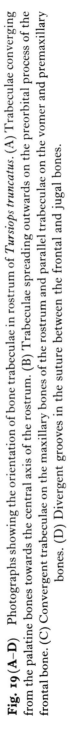

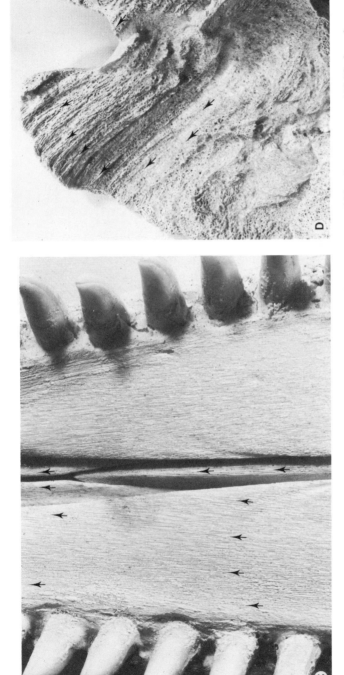

Fig. 19 (A–D) Photographs showing the orientation of bone trabeculae in rostrum of *Tursiops truncatus*. (A) Trabeculae converging from the palatine bones towards the central axis of the rostrum. (B) Trabeculae spreading outwards on the preorbital process of the frontal bone. (C) Convergent trabeculae on the maxillary bones of the rostrum and parallel trabeculae on the vomer and premaxillary bones. (D) Divergent grooves in the suture between the frontal and jugal bones.

The internarial, posterior aspects of the palatine bones, to which the two sets of palatopharyngeal muscles (MPP) are attached, are concave antero-posteriorly and elongated in a ventro-dorsal direction. Their ventral extremities are separated in the Common porpoise by approximately 2 cm, but diverge dorsally, so that their upper extremities are separated by about 5 cm.

This general form and orientation of the palatine bones is common to all odontocetes, but the lateral separation of the bones varies throughout the order. Schenkkan (1973) prepared radiographs of young and old skulls of *Phocoena* showing that the orientation of the rostral trabeculae is consistent throughout the bone.

IV. Anatomy of the Dorsal Aspect of the Head

Figures 20–24 show the anatomy of the head of a delphinid, in this case *Steno bredanensis*, but the arrangement applies to all the Delphinoidea with minor differences.

A. Superficial muscles

In Fig. 20 the panniculus adiposus or blubber has been removed except in the region of the blowhole. The first peculiarity to be noticed is the sinistral position of the blowhole. The lateral axis of the latter is set at an angle to the long axis of the head. The muscles which control the dilation of the blowhole and compression of the underlying diverticula of the nasal tract are also asymmetrical and make an angle with the long axis of the head. It is this fact that is responsible for the asymmetry of the skull. It was pointed out earlier that during the course of evolution bone becomes altered to suit changes in the anatomy of the soft parts and not vice versa. It was also pointed out that the antero-posterior asymmetry is due to the fact that during deflection of the nasal tract to the left the lateral axes of its various diverticula have remained at right angles to the vertical axis of the nasal passages and so present a dorso-ventral asymmetry as well as an antero-posteriorly skewed appearance.

There is no doubt that the muscles shown could be subdivided into smaller components based on the arrangement in terrestrial mammals, but since there has been extensive fusion of the components they are

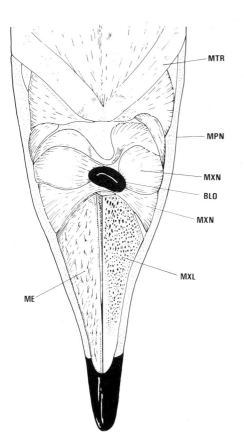

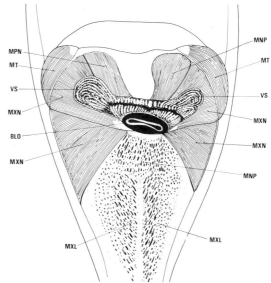

Fig. 20 Diagram of a dissection of the dorsal aspect of the head of *Steno bredanensis* after removal of blubber and the left half of the melon.
BLO = blowhole, ME = melon, MPN = procerus nasi muscle, MTR = trapezius muscle, MXL = maxillolabialis muscle, MXN = maxillonasalis muscle.

Fig. 21 Diagram of a dissection of the dorsal aspect of the head of *Steno bredanensis* after removal of the superficial muscle showing vestibular air-sacs.
BLO = blowhole, MNP = procerus nasi muscle, MPN = nasal plug muscle, MT = temporal muscle, MXL = maxillolabialis muscle, MXN = maxillonasalis muscle, VS = vestibular sac.

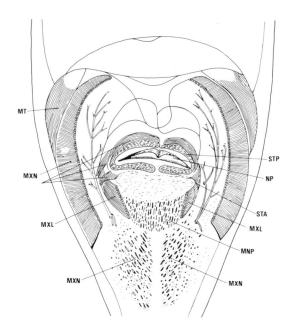

Fig. 22 Diagram of a dissection of the dorsal aspect of the head of *Steno bredanensis* showing tubular air-sacs and the functional part of the maxillolabialis muscle.
MXL = maxillolabialis muscle, MXN = maxillonasalis muscle, MT = temporal muscle, NP = nasal plug, STA = anterior tubular sac, STP = posterior tubular sac.

Fig. 23 Diagram of the dorsal aspect of the head of *Steno bredanensis* showing "connecting" or accessory sac and roof of the premaxillary sacs.
MIT = intrinsic musculature of tubular sacs, MSM = supramaxillary sac muscle, MXL = maxillolabialis muscle, NP = nasal plug, PMX = premaxilla, SCO = connecting sac, SMA = supramaxillary sac.

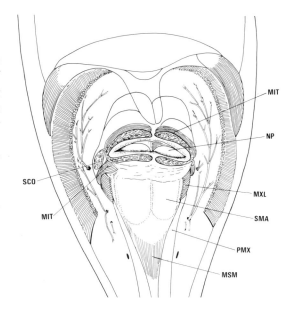

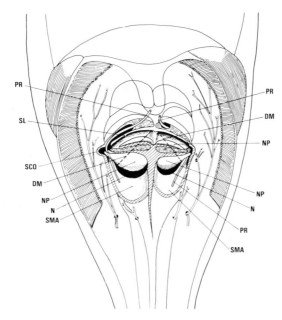

Fig. 24 Diagram of a dissection of the nasal region in *Steno bredanensis* to show the connection between the tubular sacs and the premaxillary sacs and the relationship of the nasal plugs to the nares. The figure also shows the valvular lateral extensions of the nasal plugs.

DM = diagonal membrane, N = naris, NP = nasal plug, PR = probe, SCO = connecting sac, SL = slot leading to posterior tubular sac, SMA = supra maxillary air-sac.

difficult to dissect. The blood vessels and nerves supplying the muscle mass have become much less diversified than in terrestrial mammals and it would appear that the whole system works in unison or in sequence, there being very limited provision for independent action of its various parts. Starting from the rear there are the procerus nasi muscles (MPN) which are attached to the frontal bone and draw the lips of the blowhole posteriorly. Anterior to these are two components of the maxillonasalis (MXN) which form the roof of the underlying vestibular air-sacs and serve to compress the latter. More anteriorly there are the main maxillonasalis muscles which dilate the blowhole laterally and anteriorly. Finally there is the maxillolabialis (MXL) which in cetaceans has undergone great hypertrophy and fatty degeneration to form the so-called melon (ME).

In Fig. 20 one half of the melon has been cut away to reveal the vestiges of the red muscle fibres which are attached to the maxillary bones of the skull. It should be emphasized that there are no muscles for occlusion of the blowhole, this being maintained by internal pneumatic pressure as will be seen later.

B. Vestibular air-sacs

In Fig. 21 the superficial musculature has been removed to show a pair of vestibular air-sacs lined by a wrinkled epithelium. At this level the aperture of the blowhole is much wider and tends to become divided into two sections which are the homologues of the paired nostrils of terrestrial mammals. The vestibular sacs are the homologues of the vestibula of the nostrils and anterior nasal tracts of terrestrial mammals and if the nostrils were to move forward again to their original position at the tip of the snout, the folds in the epithelial lining would partially disappear and become antero-posteriorly orientated as has happened in the secondary forward displacement of the blowhole in the Sperm-whale.

There is a small group of fibres immediately anterior to the blowhole which contains a few red muscle strands which have a sinistral orientation and have been referred to as "nasal plug muscles" and are consequently labelled MNP. It is our view that they can have limited muscular action and are another part of the degeneration of the maxillolabialis. Their main function seems to be that of preventing the nasal plugs (to be referred to shortly) from being thrust down the posterior nares under extreme hydrostatic pressure.

C. Tubular air-sacs

In Fig. 22 the dissection is in part more superficial and in part deeper than that in Fig. 23. Thus there is a small fully functional, transverse band of the maxillolabialis (MXL, the cut ends of which are shown in the diagram) that lies superficial to the "nasal plug muscles". This serves to compress the underlying premaxillary sacs and assists in the recycling of air during sonar emission. It also compresses the "connecting sacs", more appropriately named the accessory sacs by Schenkkan (1973). See also Fig. 23 (SCO).

This transverse band also presses the lateral flaps of the nasal plugs (NP) against the posterior, dorsal surfaces of the premaxillae during sonar emission.

This figure also shows a pair of U-shaped "tubular sacs" which surround the blowhole. Each has an anterior and a posterior section labelled STP and STA respectively. These are diverticula of the nasal tract and have their own intrinsic musculature labelled MIT. The upper surfaces of the nasal plugs (NP) are also shown. The arteries supplying the musculature of the blowhole can be seen emerging from foramina in the maxillary bones.

D. Premaxillary air-sacs and nasal plugs

Figure 23 shows the membranous, shield-shaped roof, the premaxillary sacs (SMA), and a mesial line dividing the area into right and left sections, the right being larger than the left. A pair of muscles labelled (MSM) lies in the interval between the premaxillary bones and have been called "premaxillary sac muscles". This name has been applied by other anatomists but the muscles can easily be homologized with the depressor septi muscles of terrestrial mammals because of their mesial attachment to the nasal septum. Similarly the muscles labelled (MIT) in this and the other diagrams have been called the "intrinsic muscles" of the blowhole but these can also be homologized with the compressor and depressor nasi of terrestrial mammals. In Fig. 24 the roof of the premaxillary sacs has been removed so that both the upper and lower surfaces of the nasal plugs have been exposed. The bony nares have also been exposed and it will be seen that the ventral surfaces of the nasal plugs fit exactly into these apertures. There are concavo-convex lateral extensions of the nasal plugs which are directed towards the accessory sacs (SCO). These flaps act as unidirectional valves which will allow air to pass upwards into the tubular sacs but will not allow air to pass downwards once the tubular air-sacs have been inflated. These valvular flaps have been regarded as the place of origin of the sonar pulses by Norris (1964, *loc. cit.*)

Incisions have been made in the posterior tubular sacs and two probes (PR) have been inserted to show the connection between the tubular sacs and the premaxillary sacs. Two "diagonal membranes" lie across the lateral regions of the bony nares deep to the nasal plugs and their mesial borders stretch from the mesial septum to a point just under the posterior extremities of the premaxillary bones. The mesial borders have been represented by a dotted line labelled (DM) in the diagram.

Green *et al.* (1980) and Ridgway *et al.* (1980) have described a "diagonal membrane muscle" and an electromyographic study has been made

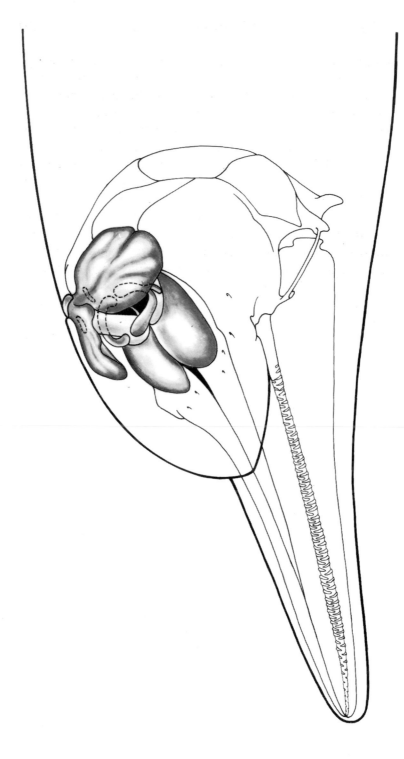

Fig. 25 Three dimensional diagram of the whole assembly of air-sacs in an inflated condition associated with the recycling of air during echolocation in a typical delphinid.

of this muscle. One can only assume that this is a case of mistaken identity as the diagonal membrane is purely valvular in character and contains no muscle whatsoever. The two membranes lie across the upper bony nares and consist entirely of elastic and white fibrous tissue. The membranes are firmly adherent to the bone of the nostrils on their anterior, posterior and lateral margins so that no significant muscular contraction is possible.

It was stated earlier that the nares had been displaced backwards throughout evolution from the front of the rostrum to the top of the head carrying with them their epithelial lining which in recent cetaceans forms the lining of the vestibular sacs. During this process the linings of the nasal sinuses have also been carried backwards and the so-called "posterior and anterior tubular sacs" can be homologized with the frontal and ethmoidal sinuses of terrestrial mammals. Figure 25 is a three dimensional diagram of the whole arrangement of air-sacs in the upper narial region of a typical delphinid and this should be consulted during the chapter on Phonation which follows.

III.   Phonation in Cetaceans

Romanenko (1976) in his review of experiments on the acoustics of cetaceans carried out in the U.S.S.R. stated:

> "For example, it has been established that the dolphin's mechanism for the emission of high-frequency pulses has a non-resonance character, refuting the hypothesis advanced many years ago by J. C. Lilly and accepted as entirely reliable until the present time, that the emission process is based on a resonance mechanism".

Purves (1966) described in detail what he thought was the site of the sonar emissions and why these were transmitted to the ambient sea water directly, without the intervention of resonance phenomena. Romanenko also goes on to state:

> "It has been confirmed that the width of the echolocation pulses emitted by a dolphin differs in certain directions. Forward-emitted pulses are approximately twice the width of the pulses emitted to the sides. The latter can clearly be regarded as elementary directly emitted pulses, while those propagating in the forward direction appear to be formed from the elementary, by means of the frontal and supramaxillary bones of the skull and soft tissues of the cranium".

Purves, in the paper cited above, domonstrated with very primitive apparatus, that the forward directed pulses also contained approximately twice the energy content of the side directed pulses—this he presumed being due to the trabecular arrangement in the frontal, maxillary and premaxillary bones of the skull. At that time the melon was not thought to be directly involved. Romanenko does not state the origin of the echolocation pulses but the experimental results seem to confirm the anatomical findings. As the hypothesis of Norris and his co-workers, stating that the lateral extremities of the nasal plugs form the origin of echolocation pulses with the melon acting as an "acoustic lens", has been given almost world-wide credence it has been decided to review the current theories of phonation in the light of recent developments.

A great deal of speculation has arisen about the mechanism of phonation, and a number of different hypotheses have appeared in the literature. Some authors, notably Kellogg (1966), Evans and Prescott (1962), Norris (1964), Evans *et al.* (1964), locate the sound-producing mechanism among the various pneumatic chambers which are to be found in the upper part of the blowhole, but others, Lawrence and Schevill (1965), believe the sound source to be laryngeal. Lilly (1962, 1964), who carried out an exhaustive analysis of cetacean noises under ideal conditions, maintains that there are at least two, and possibly three sound-producing mechanisms which may be used separately or simultaneously. In his earlier papers he is inclined to the view that these systems are located in the upper nasal chambers, but in his book "*Man and Dolphin*" he concedes that the larynx may also be involved.

When one recalls the great variety of noises which may be elicited by simply blowing air through the nasal passages of a dead cetacean, it would not be surprising to discover that an intelligent animal like a dolphin could make use of the sound producing properties of its valves and air chambers to produce airborne "humanoid" noises for communication with Man, and in certain circumstances with its own species. It is evident, however, that the so-called separate "layers" of the muscle system which control the air-sacs and valves can only be operated in sequence and not independently of one another. All the blood vessels and nerves which supply the major groups of muscles spring each from a separate main trunk and pass from layer to layer within the group, branching only within the "layer" of the muscle. This may be demonstrated by injecting the system with latex or polyester resin, whereupon the "layers" become quite inseparable owing to the numerous interconnecting vessels. The shearing stresses involved in independent action of the muscle "layers" would be bound to cause rupture of the vessels and nerves.

It is noteworthy that Ridgway *et al.* (1979) during their electromyograph studies on phonation in *Tursiops truncatus* found elicited potentials in what they term the left and right internus muscles when the animal was whistling or blowing. These muscles, which we regard as part of the maxillonasalis, are inserted into the upper part of the nasal tract above the level of the nasal plugs.

Lawrence and Schevill (1965) have demonstrated most elegantly that all the various air-sac systems are necessary for the non-muscular occlusion of the blowhole during submergence, but it can be shown that they are also necessary for the retention and recycling of air during phonation without themselves necessarily being involved in the production of sound.

This chapter is primarily concerned with sonar and it is proposed to show that phonation by the larynx is quite adequate to meet the requirements of echolocation. The normal mechanism of phonation in mammals is located in the larynx and the fact that there are no vocal chords of the conventional type in the cetacean larynx has brought about its immediate rejection as a possible sound source. However, as Negus (1949) has shown, Man is one of the few mammals which phonate by means of true "vocal chords" the majority employing modifications of the thyroarytenoid fold, and a few, the artiodactyl ungulates, of the aryepiglottic folds. It has been our experience that whatever structure is responsible for a certain function in terrestrial mammals is also responsible for that function in cetaceans, except that the structure may have undergone profound modification for operation under water. *Situated as it is, directly between the peribullary air spaces, the glottis is substantially acoustically isolated from the ears, and thus in a favourable position for the role of sound emitter. In echolocation, the voice must necessarily be mainly monitored on the echo, rather than on the emitted pulse, if the repeat-frequency is to be correctly adjusted for range. It is therefore essential that no unattenuated sound path from emitter to receiver should be available.*

Acting on the assumption that the larynx was the main sound emitter, Purves carried out experiments on the propagation of laryngeal sound through the head in fresh specimens of *Phocoena phocoena*, *Lagenorhynchus cruciger* and *Tursiops truncatus* i.e. on blunt-snouted, short-snouted and long-snouted cetaceans respectively. The experiments were so simple and striking in their results that it became a matter of routine to store in deep freeze animals that had been stranded alive and had subsequently died, for demonstration of sound propagation to students visiting the British Museum.

I. Experiments and Methods

The first experiment was carried out on a fresh specimen of the Common porpoise, *Phocoena phocoena*, which had been caught accidentally by fishermen in the North Sea. For the purpose of the experiment a variable-pitch, Galton whistle commensurate with the larynx was adjusted to 20 kHz, and attached to an air line, which was then introduced into the trachea until the body of the whistle lay in the rima glottidis with the case between the arytenoid cartilages. The whistle was considered to be necessary since the mere act of blowing air through the larynx without the whistle produced non-specific noises in many parts of the respiratory tract which could not be monitored satisfactorily. The introduction of electrical devices for the production of sound in the larynx inevitably produces capacitance effects which are distributed with equal intensity over the whole body as Mackay (1964) discovered to his advantage during his experiments on ingested micro-transmitters in *Tursiops*. It was found that an air pressure equivalent to 25 mm Hg was sufficient to produce an intense and extremely high-pitched, but audible tone, after the incision had been tightly sutured round the air line. Owing to the self-sealing action produced by inflation of the tubular air-sacs, as described by Lawrence and Schevill (1965) a sustained note could not be maintained by this method for more than 10 s, so that the lips of the blowhole had to be kept apart by intubation during a prolonged series of measurements. The sound intensity at various points on the head was found by measuring the longitudinal vibration of a suture needle inserted alternatively superficially and deeply at these points. A piezoelectric cartridge was placed so that the stylus rested on the eye of the needle at right angles to its axis, and the output of the crystal was connected to a cathode-follower, amplifier, volt-meter and oscilloscope. The lowest voltage after amplification (0·1 V) was used as the reference in recording intensity levels in db.

It was not the purpose of these experiments to measure the intensity of vibration of air in the nasal passage, but that of the small displacement-amplitude, high-energy vibrations transmitted through the body of the whistle to the cartilages of the larynx, assuming these to be further trans-mitted to the vomer and via the palatopharyngeus muscles to the maxillary bones of the skull. Such measurements required that there should be no large acoustic mismatch at any point. The vibrations of the animal tissues were, therefore, coupled directly to the piezoelectric cartridge via the suture needle.

The figures for the first experiment with *Phocoena phocoena* are given in Table 1 and Fig. 26a.

Almost immediately afterwards, a deep-frozen specimen of *Lageno-rhynchus cruciger* was received in absolutely fresh condition in January 1960 from Dr C. E. Ash, a Whaling Inspector aboard the "Balaena". The

Table 1: Showing distribution of sound intensity levels in the head of a Common porpoise, *Phocoena phocoena*, radiated from an artificial sound source in the larynx.

Horizontal plane Position	dB S	D	Sagittal plane Position	dB S	D	Lower jaw Position	dB S	D
2 cm above eye	6	6	2 cm anterior to blowhole	0·13	1·6	2 cm below eye	0·2	1·2
4 cm anterior to eye	0·78	6	6 cm anterior to blowhole	0·13	23·4	4 cm anterior to eye	0·2	0·2
8 cm anterior to eye	0·6	6	10 cm anterior to blowhole	0·13	6+	12 cm anterior to eye	0	0
12 cm anterior to eye	0·6	7·8	14 cm anterior to blowhole	0·78	20·0	16 cm anterior to eye	0	0
16 cm anterior to eye	0·13	9·4	18 cm anterior to blowhole	1·2	20	18 cm anterior to eye	0	0
20 cm anterior to eye	0	15				20 cm anterior to eye	0	0
Tip of snout	15·4	23	Tip of snout	23	23	Tip of lower jaw	0	0

S = Superficial   D = Deep   + = Premaxilla

It is important to note that the above dB figures are not absolute units of intensity as defined by the British Standards Institution for human hearing—i.e. that corresponding to an R.M.S. sound pressure of 0·0002 dyne/cm² at a frequency of 1kHz. They are derived from voltage readings above the given reference level.

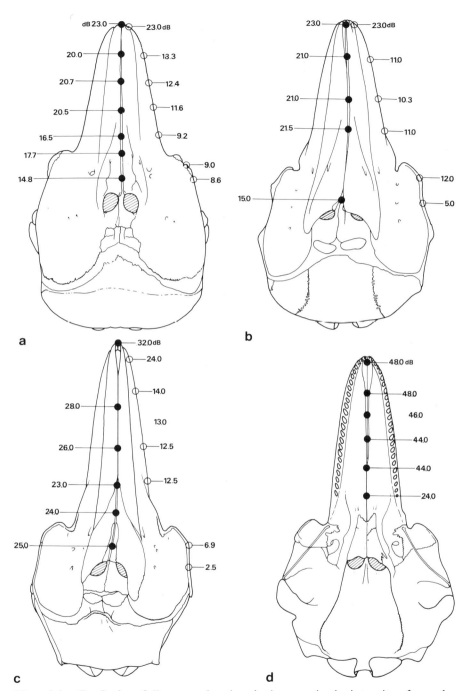

Fig. 26 (a–d) Series of diagrams showing the increase in the intensity of sound levels towards the tip of the snout and the central axis of the skull obtained from a sound source placed in the larynx of three species of cetacean: (a) *Phocoena phocoena* (see Tables 1 and 3); (b) *Cephalorhynchus cruciger* (see Table 2); (c) *Tursiops truncatus* dorsal (see Table 4); (d) *Tursiops truncatus* ventral (see Table 4).

Table 2: Showing distribution of sound intensity levels in the head of a Crucigerous dolphin, *Lagenorhynchus cruciger*, radiated from an artificial sound source in the larynx.

Horizontal plane Position	dB S	D	Sagittal plane Position	dB S	D	Lower jaw Position	dB S	D
2 cm above eye	3·3	5	4 cm anterior to blowhole	0·3	15	2 cm below eye	0	0
4 cm anterior to eye	1	12	8 cm anterior to blowhole	0	0	4 cm anterior to eye	0	0
12 cm anterior to eye	0·75	11	12 cm anterior to blowhole	0	21·5	12 cm anterior to eye	0	0
16 cm anterior to eye	0·75	10·25	16 cm anterior to blowhole	0	21	16 cm anterior to eye	0	0
32 cm anterior to eye	0·3	11	32 cm anterior to blowhole	0	21	32 cm anterior to eye	0	0
Tip of snout	23	23	Tip of snout	23	23	Tip of lower jaw	0	0

S = Superficial   D = Deep

experiment was repeated in the same way as with *Phocoena*. The results for *Lagenorhynchus cruciger* are given in Table 2 and Fig. 26b.

Inspection of the figures shows that:

(a) Maximum values were obtained from the tip of the snout, notwithstanding that this point was furthest away from the source of sound.

(b) The maximum value is the same in both species despite the fact that the length of the head of *L. cruciger* is almost double that of *P. phocoena*.

(c) The superficial values are very much lower than those measured with the needle in contact with bone, except in the regions where the soft tissues were relatively thin, i.e. in the region just anterior to the eye, the hard palate and at the tip of the snout.

(d) Relatively large signals were obtained in the vicinity of the vomer.

(e) Negligible signals were recorded on the lower jaw.

(f) Deep signals taken in the sagittal plane are of much greater strength than those taken on one side of the horizontal plane and are frequently double the signal strength.

The very low intensities recorded from the superficial structure of the head could be accounted for in a number of ways: firstly, by mechanical slip in the suture needle; secondly, by the large difference in acoustic impedance between the steel of the suture needle and the blubber of the melon; and thirdly by the fact that the longitudinal waves in the bones of the skull were arising as transverse waves in the melon. This last explanation is probably the correct one as it is a common phenomenon where bodies conducting longitudinal waves are immersed in liquids.

Some time later, whilst carrying out economic work at the Zoological Laboratory, Amsterdam, Purves was able to repeat the experiments on two more specimens of *Phocoena* and on a Bottle-nosed dolphin *Tursiops truncatus*, but no superficial readings were taken at this time. When the second experiment on *Phocoena* was performed, completely negative results were obtained, and the animal was taken away for autopsy. On removing the flesh from the skull, it was found that the rostrum had been fractured completely through its base and was detached from the remainder of the skull. Another specimen was then experimented upon and the results were almost exactly the same as those obtained in London.

The results obtained from *Tursiops* (Fig. 26c, d) were in general similar to those obtained in *Phocoena* but in this experiment a more sensitive cartridge was used, a fact which may account for the higher voltages obtained. The results for *Phocoena* and *Tursiops* are given in Tables 3 and 4.

Table 3: Showing distribution of sound intensity levels in the head of a common porpoise *Phocoena phocoena* radiated from an artificial sound source in the larynx.

Horizontal plane	dB	Sagittal plane	dB	Lower jaw	dB
2 cm above eye	11·25	2 cm anterior to blowhole	16	2 cm below eye	0
4 cm anterior to eye	11·75	4 cm anterior to blowhole	12	4 cm anterior to eye	0
8 cm anterior to eye	12·5	8 cm anterior to blowhole	16·5	8 cm anterior to eye	0
12 cm anterior to eye	15·5	12 cm anterior to blowhole	21·5	12 cm anterior to eye	0
16 cm anterior to eye	15·5	16 cm anterior to blowhole	21·5	16 cm anterior to eye	0
20 cm anterior to eye	21·75	20 cm anterior to blowhole	20·75	20 cm anterior to eye	0
Tip of snout	23·0	Tip of snout	23·0	Tip of lower jaw	0

Table 4: Showing distribution of sound intensity levels in the head of a bottle-nosed dolphin *Tursiops truncatus* radiated from an artificial sound source in the larynx.

Horizontal plane	dB	Sagittal plane	dB	Lower jaw	dB
Opposite ear	0			Below ear	0
4 cm anterior to ear	0			4 cm anterior to ear	0
3 cm above eye	12·5			3 cm below eye	1·5
4 cm anterior to eye	6·0			4 cm anterior to eye	0
8 cm anterior to eye	12·5	2 cm anterior to blowhole	25	8 cm anterior to eye	0
12 cm anterior to eye	6·0	4 cm anterior to blowhole	24	12 cm anterior to eye	0
16 cm anterior to eye	13	8 cm anterior to blowhole	23	16 cm anterior to eye	0
20 cm anterior to eye	14	12 cm anterior to blowhole	26	20 cm anterior to eye	0
24 cm anterior to eye	13·5	16 cm anterior to blowhole	28	24 cm anterior to eye	0
Tip of snout	24	Tip of snout	32	Tip of lower jaw	0
Hard palate (lateral)	dB	Hard palate (medial)			
Level with first tooth	8	Mid-line level with first tooth	24		
4 cm anterior to first tooth	12	4 cm anterior to first tooth	44		
8 cm anterior to first tooth	12	8 cm anterior to first tooth	44		
12 cm anterior to first tooth	22·5	12 cm anterior to first tooth	46		
16 cm anterior to first tooth	29	16 end of vomer	48		
Tip of snout	23	Between maxillae	48		

Diagrams relating to Tables 1–4 are shown in Fig. 26 (a–d).

The figures of *Tursiops truncatus* are particularly noteworthy because on this occasion an attempt was made to take signals from the vicinity of the external auditory meatus with completely negative results. This implies that the external meatus is also screened acoustically from sound generated in the larynx. The signals taken from the hard palate inside the mouth are also of particular interest since those taken along the mid-line of the palate are frequently more than double those taken at the side of the palate. One cannot avoid the conclusion that the signal strength along the mid-line palate represents the summation of the signal strengths from the two sides of the rostrum.

Before going on to discuss the function of the various parts of the vocal and nasal tracts it should be pointed out that some workers tend to regard the epiglottic spout as an organ distinct from the larynx. The epiglottic spout is an intrinsic part of the larynx and has been formed by great hypertrophy and elongation of the epiglottic and cuneiform cartilages. The latter are relatively minute in Man and in most terrestrial mammals and in cetaceans are permanently fused to the arytenoid cartilages. Therefore when the larynx is referred to in this paper it is meant to include the epiglottic spout.

In August 1963, one of us (P.E.P.) spent a week at the Communications Research Institute, Miami, U.S.A., where experimental work is carried out on captive specimens of *Tursiops truncatus*. In one of the laboratories, the animals were kept in large fibre-glass tanks, each having a narrow Plexiglass annex as shown in Fig. 27. When the tanks were first installed in 1961, the animals were driven into these annexes for the purpose of making psychological and cathode-ray encephalograph studies, but since that time had been in the habit of entering them voluntarily, spending a great deal of time observing the activities of people in the laboratory. They were also able voluntarily to back out of the annexes into the main tank if they so wished. This arrangement enabled sound intensity experiments to be carried out without having to restrain the animals in any way and without the use of hydrophones.

A pair of flat-faced barium titanate transducers 1 cm square were clamped to the walls of the annex (Fig. 27) and coupled acoustically to the Plexiglass by coating with a thin film of petroleum jelly. The transducers were then connected by long, coaxial cables to two identical band-pass filters, preamplifiers, and a double-beamed oscilloscope which were enclosed in an instrument booth out of sight of the animal. One transducer was fixed to the end of the annex, and coupled to the upper trace of the

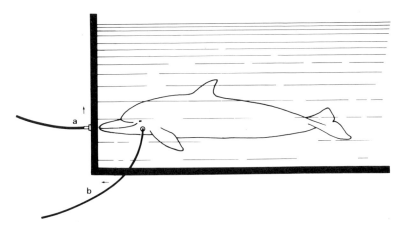

Fig. 27 (a–b) Diagram of a living specimen of *Tursiops truncatus* having voluntarily entered the plexiglass annex to an aquarium using echolocation: (a) barium titanate transducer on the front of the annex level with the tip of the snout. (b) identical transducer placed at the side of the annex approximately opposite the larynx.

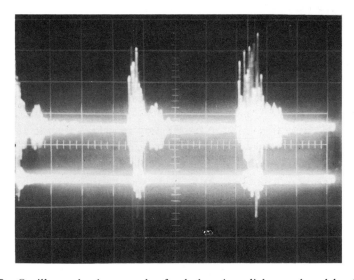

Fig. 28 Oscillograph photograph of echolocation clicks produced by *Tursiops truncatus*. Upper trace from transducer (a) in Fig. 27, lower trace from transducer (b).

oscilloscope and the other to the side, at a point approximately in line with the glottis and posterior nares of the animal when it was fully in the annex. This transducer was coupled to the lower trace of the oscilloscope.

The mere activity of fixing the apparatus was sufficient to excite the animal's curiosity and it would enter the annex emitting bursts of sonar activity at the same time. The emitted sounds were of three main types, a clicking pulse of variable repeat frequency, short or sustained whistles, and loud squawks. During the last two types of phonation, a considerable amount of noise was emitted from the blowhole into the air, if the former happened to be out of water at the time. It should be pointed out at this stage that it is the latter phenomenon which may have led to the assumption by a number of authors that the noise emitter is the blowhole mechanism. However, it is easy to be misled in this respect. In the laboratory experiments on the dead animals, previously described, the sound of the whistle seemed to come from the blowhole, notwithstanding that the whistle was located deeply in the larynx. It is significant also that the dolphins could emit all these noises with equal facility, though with less intensity, when the blowhole was closed and the head completely submerged.

The filters were set to pass frequencies above 30 kHz since the transducers were relatively insensitive below this level. With the oscilloscope trace set at 10 Hz when the animal was just entering the annex at a distance of 1·5 m approximately, the repeat frequency was 80 pps. The amplitude on the upper trace was greater than that on the lower, the measured voltages being 0·12 V and 0·04 V respectively; difference 9·6 dB approximately, the effect of a greater amplitude on the upper trace might have been expected irrespective of any directionality of the pulse owing to the channelizing effect of the annex and the angle of reflection at the side transducer. Figure 28 shows the trace set at 100 Hz when the animal was nearly touching the end of the annex; the repeat frequency was 300 pps and the voltage strength on the upper and lower traces 0·28 V and 0·09 V respectively, difference 9·8 dB. Since the camera exposure was 0·04 s each pulse shown on the photograph represents a group of approximately four clicks superimposed. As might be expected at this repeat frequency, individual clicks could not be distinguished by ear, the general effect being that of a fairly high-pitched squawk. In this last position the distance from the sound-producing mechanism (whether it be the larynx or the nasal passages), to the end transducer, was at least four times that to the side transducer, and yet the amplitude of the sound at the end transducer was approximately 9 dB above that at the side transducer. This result showed that all frequency components above 30 kHz were being beamed predominantly forward along the rostrum. The absence on the lower trace of the double pulse should be noted. A great number of other measurements were taken at this time but as the polaroid records have now been destroyed they cannot be reproduced here. It should be stated, however, that there was a marked drop in signal strength when transducer A (Fig. 27) was raised above the level of the rostrum, and an increase in intensity as transducer B was moved forward (see Purves, 1966). The results were in general conformity with those obtained from the dead animals previously described.

In a paper by Diercks *et al.* (1971) on "*Recording and Analysis of Dolphin Echolocation Signals*" in which transducers were applied at various points on the head of a Bottle-nosed dolphin, *Tursiops truncatus*, the authors found that the amplitude of the signals obtained from the rostrum was significantly greater than that from any other part of the head. But then they go on to state:

> "Assuming straight-line transmission through soft tissue the time differences between wave forms shown in Fig. 5 place the click generator at the location of the nasal plugs in the nares, a site also implicated by anatomical evidence".

If they had located the click generator at the tip of the glottis as we do, they would have found that the time differences exactly corresponded with the distance travelled in the structures above the glottis. In any event their results are very similar to those described by us and it is obvious that the signal most useful in echolocation would be that of the greatest amplitude.

II. Function of the Larynx

In considering the function of the cetacean larynx it is necessary to take into account the fundamental difference between the mode of propagation of sound in air and water. These differences in the physical properties of sound propagation in the two media were seen to have been of great importance in the interpretation of the modification of structures in the middle ear (see Fraser and Purves, 1960), and it is not surprising that they are also relevant to the understanding of modifications in the larynx.

In all terrestrial mammals, phonation involves the vibration of one or more of a variety of structures, such as vocal folds, "vocal chords", thyroarytenoid folds, thyroaryepiglottic folds, etc., which in some cases operate in conjunction with resonant cavities in the mouth and nasal cavities (Negus, 1949). In every case, air disturbances set up in the vicinity of these structures are propagated directly to the ear of the hearer through the air in the atmosphere. Thus there is no discontinuity of medium between transmitter and receiver, and consequently no loss of energy, except that which accrues from attenuation. In the cetacean, air disturbances of a similar nature would have to be transmitted through solid tissues and the surrounding sea water before reaching the ear of another cetacean. It is well known that where a gas-liquid interface is infinite or semi-infinite in area with respect to the wavelength of an emitted vibration, there is over 99% reflection of energy at the interface, whether the vibration is initiated in the gaseous or in the liquid medium. This is perhaps the main reason why we believe that *air* vibrations in the larynx and nasal passages of cetaceans are relatively unimportant, and that phonation does not take place by the operation of any structure involving simple harmonic motion. The general principle seems to involve the mechanical vibrations of the relatively heavy structures of the glottis which are transmitted to the pharyngeal muscles and thence to the bones of the rostrum, and finally to the sea water without change of medium. Other reasons are concerned with the recorded frequencies and qualities of sounds emitted from living cetaceans and of sound produced artificially in larynges in the laboratory.

It is generally agreed that in Man, the higher pitch of the voice in females is associated with the shorter "vocal chords". In the small head register of the soprano, the notes are produced by vibrations of only the inner margins of the chords, with the vocal chink being reduced to a small anterior aperture which becomes smaller as the pitch rises. McKendrick, according to Burns (1921), has shown that the limits of performance of the human voice range from FL (85 Hz) in the bass to G4 (768 Hz) in the soprano. The larynx of the Bottle-nosed dolphin, *Tursiops truncatus*, is about five times the size of that of Man, and yet the frequencies of the vibrations emitted by these animals lie between the range of 5 Hz and 170 kHz (Kellogg, 1961; Kellogg *et al.*, 1953).

The noises emitted by toothed cetaceans have been variously described as being like "the sound of musical glasses when badly played", "a rusty hinge", "creaking gate", "pneumatic hammer", whistling, chirping, clicking, etc. Kellogg *et al.* (*loc. cit.*) state "one is similarly reminded of the Bronx cheer made by blowing air through tightly pressed human lips". All these noises belong to a particular class of vibration, which Van der Pol (1926) described as "relaxation oscillations". In all these examples, the frequency is not dependent on the customary ratio of elasticity and mass, but is controlled by some form of resistance, which, on reaching a certain critical value, suddenly relaxes, builds up again and so on. From mathematical considerations and analogous electrical circuits, Van der Pol (*loc. cit*) has deduced that the periodic time of a relaxation is typified by the equation:

$$T \text{ rel} = 1.61 \ r/s$$

The fundamental period of the vibration is, apart from the numerical constant, defined by a quantity involving resistence (r) and elastic forces (s) only. Summarizing the properties of relaxation oscillations, Van der Pol states:

(a) "their time period is determined by a time constant or relaxation time;
(b) their wave form deviates considerably from the sinusoidal curve, and, as very steep parts occur, many higher harmonics of great amplitude are present;
(c) a small periodic force can easily force the relaxation system into step with it (automatic synchronisation, even on subharmonics) while under these circumstances;
(d) their amplitude is hardly influenced at all".

It is proposed to show that all these properties are relevant to the production of directional beams of sound initiated in the larynx of toothed cetaceans.

In terrestrial mammals the arytenoid cartilages are normally approximated during phonation by rotating mesially on the cricoid cartilages. The rima glottidis is then further constricted by controlled contraction of the thyroarytenoid muscles (Negus, 1949). In Man, this involves partial approximation of the "vocal chords" or thyroarytenoid folds and it was once thought that when air was passed through the "chords" they would vibrate harmonically at a pitch determined by the degree of contraction of the muscles. The expression "vocal chords" has been used throughout this paper in inverted commas since there are no vocal "chords" in the mammalian larynx. The word "chord" is more strictly applicable to a stringed instrument in which simple harmonic motion is involved. In the Chiroptera the so-called "vocal chords" have been completely destroyed by thermocauterization without any impairment of the echolocating capabilities of these animals (Motta, 1959).

In a remarkable series of experiments, Husson (1962, 1963) demonstrated that the action of a single vibration of the "vocal chords" could be divided into three well-defined phases. An opening and a closing phase followed by a resting phase during which the current of air was stopped. During the closing phase which can last for less than 0·005 s, the air current receives a shock wave which raises its velocity to several hundreds of metres per second—frequently to above the speed of sound.

The sound produced at the level of the glottis is not a simple harmonic vibration but a discontinuous series of pressure waves charged with harmonic components up to ten times the fundamental frequency of vibration of the "chords".

That the discontinuity phenomenon occurs during normal phonation required experimental proof but it may have been inferred from the fact that most humans can reduce a musical tone in the larynx to a train of intermittent clicks as few as ten or less per second by strongly constricting the glottis and allowing the air to escape very slowly. During this type of phonation the breath is practically held. Husson has shown that although the volume of air expelled during each excursion of the vocal chords amounts to no more than 1 to 2 cm^3, the kinetic energy produced is very considerable. He gives the energy released per unit as:

$$e = 1/2 \ dm \ V^2 = 1/2 \ pPSV^3 \ dt$$

where dm is the mass expelled, p its volume density, P the pressure, S the surface of the glottic opening at the instant under consideration and V is the velocity of the air at the same instant. If V attains a figure of 200 m/s as for instance in the singing voice, V^3 will be in the order of 10^{22} G.C.S. units. Husson (1963) has calculated that in the bat, in which the increment of air is only a few cubic mm but of which the velocity is a little less than 330 m/sec, the energy is approximately 10^9 G.C.S. units.

According to Husson most of the energy of the high-frequency components of the voice is absorbed by the walls of the supra-glottal air cavities by the formation of eddy currents and is used in impedance matching and other homeostatic functions. From the observed phenomena, Husson defines the role of the larynx during phonation as that of injecting large amounts of energy into the supra-glottal cavities. It will be seen that in principle the vibrations of the glottis are very similar to the relaxation oscillation described by Van der Pol.

Kacprowski (1977) in his paper on *"Physical Models of the Larynx source"* has also drawn attention to the pulsed character of human phonation and carried out a detailed mathematical analysis of the function of the glottis. It is interesting to note the reason given for this research in relation to our remarks in the Preface and to work on dolphins carried out in the United States. Acoustic diagnostic methods in laryngology and phoniatry, which are based, generally speaking, on an analysis of the information content of the speech signal as the final and natural output from the speech-organs, have recently become commonly used in clinical practice. They not only assist, but in some cases are even superior to the prior classical methods, e.g. laryngoscopy, stroboscopy, electromyography and radiography etc., since:

—they are used under normal conditions of phonation and articulation,
—they neither need any surgical tool or foreign substances to be introduced into the speech organ, nor have to be aided by any dangerous and painful intrusive procedures.

In the toothed cetacean full approximation of the arytenoids at the level of the thyroarytenoid muscle is made impossible partly because of the extensive attachment of these cartilages to the lateral wings of the cricoid, and partly because of their fusion to the cuneiforms which always become approximated in advance of the arytenoids, thus preventing full approximation of the latter. Contraction of the thyroarytenoid muscle, (Figs 12, 15: MTA) draws the fused arytenoid and cuneiform cartilages (CAR, CUC) into the epiglottic trough (EPG) and intrudes the median epiglottic fold (F) into the interarytenoid space. The anterior borders of the arytenoids then being locked in the epiglottic trough are prevented from moving laterally and contraction of the interarytenoideus muscle (MIA) brings about constriction of the rima glottidis. In this condition the whole epiglottis spout becomes divided into three narrow channels and below the level of the base of the epiglottis, the laryngeal air-sacs (LAS) become strongly compressed by the thyroarytenoid muscles.

The whole action described above can be simulated on the resected larynx of a dolphin by tightly encircling the epiglottic spout with the thumb and forefinger and by pressing together with the other hand the arytenoid and thyroid cartilages. If air is then blown through the trachea the thyroid part of the *larynx will expand to approximately twice its resting diameter due to inflation of the laryngeal air-sacs*, the wings of the thyroid cartilage being simultaneously pressed apart. If prevented from doing so by manual compression, the air escapes at the distal end of the spout in the form of three narrow jets. The median jet, being flanked by the stiff, cuneiform cartilages produces a high-pitched whistle, whilst the two lateral jets, being surrounded by the more flaccid aryepiglottic folds, produce trains or bursts of staccato clicks. Both types of sound are composed of series of relaxation oscillations.

In principle, the mechanism of sound production is similar to that described by Husson for the vocal chords of Man except that the ary-epiglottic folds are called into play instead of the thyroarytenoid folds. This substitution, together with the involvement of accessory laryngeal air-sacs is fairly common in terrestrial mammals (Negus, 1949). The main difference, of supreme importance, is the formation of a triple sound source due to the presence, unique among mammals, of a dividing septum in the larynx.

When the epiglottic spout is opened by contraction of the cricoarytenoid and cricothyroid muscles, the anterior borders of the arytenoid and cunei-form cartilages are pulled out of the epiglottic trough and are thus free to separate. In these circumstances, contraction of the interarytenoid muscle pulls the anterior borders of the arytenoids and cuneiforms apart, the rima glottidis widens and the whole epiglottic spout assumes a cylindrical shape of considerable diameter. This would be the normal condition during respiration, but an indefinite series of intermediate conditions could be assumed for the recycling of air during phonation. It should be noted that if air is blown through the larynx without strong constriction of the epiglottic spout, the organ tends to assume the cylindrical shape and no sound other than a hissing noise is produced. In the experiments described by Evans and Prescott (1962) there was no provision for laryngeal constriction and it must be concluded that such sound as was produced and measured must have been generated by the passage of air through the upper nasal plugs and valves so there is little wonder that the intensity field was asymmetrical.

During submergence the blowhole is closed, the thorax is collapsed and the bronchiolar sphincter muscles are presumably closed. In such conditions there would be positive pressure in the respiratory passages, and the laryngeal air-sacs would be expanded within the walls of the thyroid cartilage. Contraction of the thyroarytenoid muscles (Fig. 14: MTA) would act in compressing the air-sacs on each side of the median folds and causing jets of air to be ejected from each aryepiglottic fold. If the interarytenoideus muscles were contracted whilst the thyroarytenoid muscles were also contracted, constriction of the rima glottidis would occur and a single jet of air would be ejected from the median aperture of the epiglottic spout.

Through the mechanisms described above it would appear that there are arrangements for the production of either a double or a single jet of air. It may also occur that in some circumstances a triple jet of air is produced. That these jets of air probably escape in the form of relaxation oscillations will be seen in considering the action of the nasopharynx.

III. Function of the Nasopharynx

Since the upper portion of the epiglottic spout is devoid of muscle it is considered that constriction of the apex of the glottis takes place through the agency of the palatopharyngeal sphincter and the arcus palatinus. In view of the known mechanism for laryngeal constriction of Man, it would appear that some entirely new structure is being invoked for a part of this function in cetaceans. However, Man is the only mammal which does not possess a larynx which is normally intra-narial. During loud phonation in most mammals, the neck is stretched and the glottis disengaged from the posterior nares, but in nearly all there is a quieter (generally higher pitched) intra-narial phonation, during which this does not occur, when the mouth is closed and the sound is emitted through the nostrils as, for instance, in the nasal whining of the dog. The role of the arcus palatinus during this type of phonation seems to be quite unknown, but the mere tonus of the muscles must, to some degree, modify the operation of the intrinsic muscles of the larynx. For reasons concerned with monitoring already referred to in the text, it is believed that during submergence, phonation in cetaceans is entirely intra-narial. By strong constriction of the apex of the glottis by the palatopharyngeal sphincter and arcus palatinus, powerful resistance would be built up against the jets of air described in the last section until the air escaped as trains of relaxation oscillations. It is obvious that the recurrence frequency of such oscillations could be controlled over a wide range to produce high-pitched, apparently pure tones, squawks, or staccato clicks. In every case the total wave form would not be sinusoidal but very steep-sided and the pulses would contain many harmonics of high intensity and frequency.

A photograph (by the Schlieren method) of a high velocity jet of air from a circular opening is shown in Fig. 29 where it is clearly seen that the emergent jet is divided into sections of nearly equal length. Kruger and Schmidtke (1919) have shown that the frequency of the tones given out by air or other gases issuing from a small, circular jet conforms with the relation $= kv/D$ where v is the velocity of the jet, D is the diameter of the opening, and k is a constant equal to 0·045 approximately for various mixtures of air, CO_2 and oxygen. Thus if, during each relaxation oscillation, the slits or jets at the apex of the epiglottic spout vary in diameter

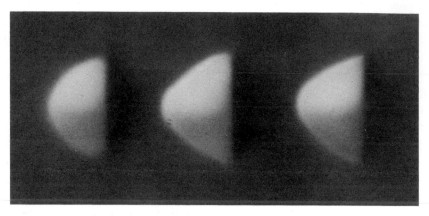

Fig. 29 Schlieren photograph showing the structure of a high-velocity air-stream from Hartmann and Trolles' air-jet generator.

between 1·0 and 0·1 mm as seems probable from observation of resected larynges, and the velocity of the jets reaches $0·333 \times 10^5$ cm/s i.e. slightly above the speed of sound (Husson, 1963), each pulse will contain frequencies between 15 and 150 kHz. This calculation is in good agreement with the frequency analysis of the echolocating pulse of *Tursiops* made by Kellogg (1961). With greater jet diameters and greater jet velocities the frequency spectrum will exceed these limits. Such frequency spectra are commonly referred to as "white noise" but the wave form of a fraction of a pulse may be revealed by the oscilloscope. In a recent study Mackay and Liaw (1981) used a modified commercial foetal heart monitor to determine the source of the dolphins echolocating pulses. They stated that the "larynx does not vibrate". Presumably the authors meant that there was no measurable vibration. We have found that the vibrations of a barium titanate transducer can be passed through the entire body at 100 kHz without gross attenuation.

If the sound source for echolocation in the cetaceans is produced by high-velocity air jets at the apex of the glottis as postulated, how then are these vibrations transmitted to the bones of the rostrum and thence to the sea water with minimum loss of energy?

As previously stated, the ventral and lateral aspects of the aryepiglottic folds at the apex of the glottis lie in a deep recess (Fig. 30) in the palato-pharyngeal sphincter, and the dorsal surface of the fused arytenoid and cuneiform cartilage makes contact with the superior constrictor muscle which is attached to the sphenoid bone at the base of the skull. The sphenoid bone is fused to the mesethmoid and vomer. The direction of the openings of the mucous ducts on the postero-dorsal aspects of the palatopharyngeal muscles is such that, if the palatopharyngeal sphincter were contracted, the recess would normally be full of mucus and consequently the apex of the glottis would be acoustically coupled to the surrounding structures. Any vibration of the cartilages and folds would therefore be transmitted to the palatopharyngeal muscles with very little loss of energy.

It is well known that cavitation, with the formation of gas bubbles, occurs when solid surfaces, immersed in liquids, vibrate with sufficiently high velocity. If the liquid contains dissolved gas, this will come out of solution when the total negative pressure at the vibrating surface in contact with the liquid falls below the saturation pressure of the gas in the liquid. Since there would normally be a great deal of dissolved gas in the mucous coating of the posterior nares after surface breathing, cavitation at the palatopharyngeal sphincter would be almost bound to occur. The presence of gas bubbles, if allowed to remain at the sphincter, would seriously interfere with the transmission of sound. It would therefore appear to be essential that some mechanism be provided for the elimination of bubbles at this point. It is for this reason that we believe that a copious flow of degassed mucus is poured over the sphincter from the mucous ducts which are so numerous and evenly distributed.

Since the pressure on the laryngeal air-sacs could be presumed to be symmetrical about the median septum, the two trains of oscillations from each aryepiglottic fold would be identical in phase and frequency. Decussation of the muscles of the palatopharyngeal sphincter on each side with those of the other, would provide for the phenomenon of synchronization described in Van der Pol's third property of relaxation oscillations.

Fig. 30 Bisected larynx of the Chinese finless porpoise, *Neophocaena asiae-orientalis*, showing the protrusion of the epiglottic spout through the palato-pharyngeal sphincter into the palatopharyngeal recess and its copious supply of mucous from numerous mucous ducts.

The fourth property of relaxation oscillations would ensure that the two lateral trains of vibrations were equal in amplitude. In the formula T $rel = Kr/s$ the value r can be regarded as being equivalent to the force exerted by the palatopharyngeal sphincter, and the value s, equivalent to the difference between the inherent elasticity of the cartilages and muscles, and that of the air in the larynx. Since the value s is a constant, the period T rel could be directly controlled by regulation of the palatopharyngeal sphincter. A steady pressure could be maintained on each side of the larynx by simultaneous contraction of the two air-sac systems, through the operation of the thyroarytenoid muscles.

Schenkkan (1973) states:

> "One of the most important characteristics of the sonar apparatus in cetaceans would be the effective range of its low frequency component, so it would be interesting to calculate the range of a train of pulses in a typical odontocete such as *Tursiops truncatus*".

The intrinsic musculature of the larynx of an adult North Atlantic Bottlenosed dolphin weighed 400 g, and cetacean muscle is capable of a short burst of energy of 4–5 kg m/s/kg muscle (Gray, 1936; Purves *et al.*, in preparation). Taking the larger figure, this is equivalent to 20 W approximately for 400 g of muscle. A Fessenden echosounder working at 50% efficiency and consuming 1kW had a range of 32 km (Wood, 1955). If the two types of oscillators are comparable at all, the larynx of *Tursiops* should be capable of producing pulses with a range of 640 m assuming a working efficiency of 50%. It may be purely coincidental that, as previously stated, the North Atlantic Bottle-nosed dolphin is capable of making dives up to about 600 m. It is conceivable that for a dolphin it would be important to perceive the availability of food at this depth before it dives since respiratory requirements will limit the duration of such a dive to about 10 min.

If the transmission of sound energy from the apex of the glottis (assuming the latter to be acoustically coupled to the palatopharyngeal muscles in the manner described above) is considered in conjunction with the distribution of air spaces it is not difficult to see why the maximum energy is found along the central axis of the rostrum.

As they ascend the posterior nares, the two palatopharyngeal muscle masses are enveloped ventrally and laterally by extensions of the pterygoid air sinuses (see Fig. 8: PTS), and dorsally by the lumen of the nares themselves. Any sound energy imparted to the two muscles by the two sides of the apex of the glottis would undergo total reflection at the air interface and could therefore only travel antero-dorsally along the muscle fibres towards their insertion on the posterior aspects of the palatine and maxillary bone (PAL, MX). Here the acoustic coupling must be very nearly perfect since the fascial extension of the muscle fibres and sheath enter the matrix on the bone and follow the antero-posterior orientation of the trabeculae in the form of long, dendritic processes. The excellence of this coupling can be demonstrated by the use of acoustic probes in the manner described by Fraser and Purves (1960) and Purves and Van Utrecht (1963). The sound energy of the larynx is thereby distributed to the bones of the rostrum as a *double source*.

Assuming the thyroarytenoid muscles to remain contracted, a single jet of air could be induced by contraction of the interarytenoid muscle (see Figs 12, 15: MIA). Such an air jet would escape as a train of relaxation oscillations through the minute aperture between the cuneiform cartilages which make contact with the roof of the nasopharynx (Figs 12, 15: CUC). Vibrations initiated in the cuneiforms would be transmitted to the base of the skull in the interval between the two air-sac systems as a single train of relaxation oscillations. From this point they would be transmitted to the mesethmoid and vomer as a single source. The arrangement described above provides for either a double or a single source or both simultaneously.

One of the most important theorems applicable to sound waves is Huyghen's Principle of Superposition. On this principle the resultant displacement of a particle of the medium through which two or more trains of waves are passing is obtained by the vector addition of the separate displacements due to each wave train independently. This principle is also applicable to velocities and accelerations but not to the squares of these quantities. Thus, two periodic vibrations of the same frequency, of amplitudes a and b and phase difference ε, combine to form a periodic vibration of amplitude $(a^2 + b^2 + 2\,ab \cos \varepsilon)^{\frac{1}{2}}$. If the amplitudes are equal $(a=b)$ and the phases are the same as $\varepsilon = 0$, superposition gives a vibration of double amplitude $2a$; but if the phases are opposed $\varepsilon = 180°$ the resultant amplitude is zero. In the more general case, the amplitude may vary between $(a + b)$ and $(a - b)$ according to the phase difference ε. Such superposition of vibrations is more commonly referred to as interference. In the cetacean, the bony rostrum remains divided into three sections throughout life since the two maxillary bones never fuse at the intermaxillary suture and the vomer, although fused at its proximal end to the base of the skull, lies in a channel formed by the mesial borders of the maxillary bones but never unites with them. In this respect the Cetacea are unique among the Mammalia. However, as there is no air-cavity separating the bones, sound can be conducted from one side of the rostrum to the other, though somewhat imperfectly, due to the lower acoustic impedance of the cartilage of the vomer.

If, as postulated, the two sides of the base of the rostrum constitute a double source of identical phase and frequency, the sound intensity in the sagittal plane of the head will always be at a maximum, since all the disturbances from the various elementary areas constituting the two sound sources will arrive at this plane in the same phase. If the vomer is also conducting sound the effect will be greatly enhanced. Outside the limits of the skull where the two sound fields overlap the phenomenon of inter-ference will occur, so that in directions inclined to the sagittal plane the intensity will be less, diminishing steadily to zero when the difference in distance between the nearest and farthest elements of the double source is rather more than half a wavelength. In a direction still more inclined the sound will increase again to an intensity 0·017 of that on the sagittal plane, passing through successive zero and diminishing maxima values as the inclination increases. The angle θ at which the first silence would occur, i.e. the semi-angle which would delimit the primary beam would be $\sin^{-1} 0·61\lambda/R$. Thus the primary or central beam would be confined to an area of small angle when the value R of the maximum distance between the sound sources was large compared with the wavelength of the sound emitted, i.e. the directionality would be sharp when the frequency was high. The polar distribution of amplitude and intensity after Stenzel (1927) is shown in Fig. 31.

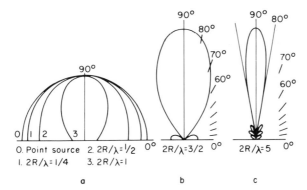

Fig. 31 (a–c) Polar diagrams showing variations in sound fields at various frequencies (after Stenzel, 1927).

The interference between two independent trains of high-frequency waves in water is demonstrated in Fig. 32.

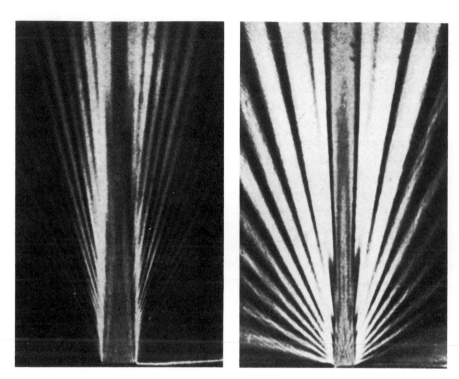

Fig. 32   Supersonic beams of sound showing primary and secondaries (after Osterhammel, 1941).

IV. Function of the Skull as a Transducer

In the previous section it was shown from anatomical and theoretical considerations why sound generated in the larynx should be concentrated along the sagittal plane of the head. Tables 1 to 4, however, show that there is a further concentration of energy at the anterior extremity of the rostrum. These tables of sound intensity measurement taken inside the mouth on the hard palate are particularly noteworthy since they show that the component bones of the rostrum tend to act as transducers concentrating energy in the anterior direction. Sound levels in dB taken in the mid-line 4 cm anterior to the first tooth, with the needle penetrating the vomer and its cartilage, are rather more than three times those taken at the same level on the maxilla. Both sets of readings increase in value towards the anterior extremity of the bone and then show a sharp drop in the soft tissues at the tip of the snout. This final drop in intensity was probably due to some form of discontinuity such as decomposition, involving the presence of gas bubbles.

From the level of the first tooth to the anterior extremity of the rostrum there is an increase in intensity in the maxilla of 21 dB and in the vomer of 24 dB. This concentration anteriorly is mainly due to the general shape of the bones of the rostrum but is probably greatly assisted by the orientation of the bony trabeculae. The vomer and its cartilage takes the form of a long, tapering rod decreasing in diameter from about 2 cm to a few mm with the trabeculae of the bone orientated antero-posteriorly in conformity with the shape of the rod. It has been suggested (not published) that the cartilage of the vomer constitutes the main transducer in the cetacean skull, transmitting longitudinal extensions of the pterygoid air-sac system on each side of the vomer on the ventral aspect of the maxillary bones. It should be borne in mind, however, that such longitudinal vibrations as are transmitted by the vomer would have directional properties only within the confines of the skull. The diameter of the tip of the vomer is so small that it would act as a point source for the whole frequency spectrum of sounds known to be emitted by cetaceans. The resultant sound field in the ambient water would be spherical and therefore non-directional. It is conceivable that this property is made use of in long-range echolocation.

So far it has been assumed that the calcarious elements of the bones of the skull transmit the sound energy but this is not necessarily the case. There is a great amount of collagen and other proteins in the bones of the cetacean skull and if these are transmitting the sound instead of the calcareous elements there would be no substantial acoustic mismatch of any kind between the vibrations of the cartilages of the larynx and those transmitted to the sea water. The radiating trabeculae of the bones of the skull would then be acting as an intricate system of "wave-guides" transmitting longitudinal sound waves to the ambient water.

In support of the theory of phonation propounded above it may be stated that if the sound produced by a small artificial whistle with relatively low air pressure in the larynx is distributed in the manner shown by Tables 1–4 and Fig. 26 (a–d), then a similar distribution of greater intensity would occur if any sound whatever were produced by the larynx in cetaceans.

Evans *et al.* (1964) and a number of other authors have suggested that the frontal aspect of the skull in toothed cetaceans including the great Sperm-whale *Physeter macrocephalus* acts as a parabolic mirror projecting the sonar pulses forward. It should be pointed out that from the engineering point of view no cetacean skull is parabolic in form and in the sub-family Monodontoidea, the front of the skull is actually convex!

The Huyghen's construction for special waves diverging from the focus is indicated graphically in Fig. 33. The aberration of the spherical and the superiority of the parabolic reflector as a means of producing *plane* waves is clearly shown.

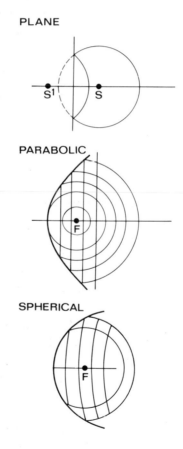

Fig. 33   The Huyghens' construction for spherical sound waves diverging from the focus of plane and curved surfaces is shown.

This is only possible when the dimensions of the reflector are large compared with the wavelength. Now a single pulse, strictly speaking, has no wavelength. The basis of comparison in such a case is the thickness or envelope of the pulse which, as the diagrams show clearly, is small compared with the diameter of the mirrors. It is also only possible when the source is close to the reflector. Neither construction is applicable to the Sperm-whale if the sound source is located at the "museau de singe" as Norris and Harvey (1972) suggest since the latter organ is located remote from the skull at the tip of the snout. It is important to draw attention to a diagram (Fig. 2) by Norris showing reflection and refraction in the head of a typical dolphin. In this diagram, which has been widely circulated in popular books throughout the world, the sound source is located in the upper narial region in the vicinity of the right nasal plug. The sound paths do not obey one of the most fundamental laws of acoustics which states that "the angle of a sound wave from a surface equals the angle of reflection incidence" as with light waves. Nor does the author indicate the large energy losses that occur at each interface owing to the differences in the acoustic impedance of the substances involved. When a laryngeal source is assumed there are no large acoustic mismatches and no multiple reflections of this type.

V. Role of the Melon

In recent, toothed cetaceans, the external apertures of the nostrils have become fused into a single "blowhole" situated on top of the head and slightly displaced towards the left. In the great baleen whales, the Mysticeti, the paired condition of the nostrils has been retained and there is no deflection towards the left.

Thus in toothed cetaceans the whole of the area between the tip of the snout and the blowhole is the equivalent of the upper lip of terrestrial mammals and it is in this light that both the anatomy and function of the skull and soft parts should be regarded. The maxillolabialis muscle has become greatly hypertrophied and in late foetal life undergone extensive fatty replacement to form the bulbous mass of fat known as the "melon".

So far, only the propagation of sound in the horizontal plane has been considered but it has been postulated that the melon acts as an "acoustic" lens for the production of directional beams of sound. In an extensive series of papers only one of which will be referred to here, Varanasi et al. (1975)

have discovered large amounts of the triglycerides of isovaleric acid in the melon (78·8%). They have also found similarly large amounts (67·4%) in the fat which covers the labial and mesial aspects of the lower jaw. This condition appears to be unique amongst the Mammalia and, as an acoustic function has been assigned to these deposits of fat by Norris (1964), they are now commonly referred to as "acoustic tissue". It should be pointed out however, that all these deposits of fat are derived from fatty replacement of muscle fibre which consists mainly of protein. This fatty replacement is correlated with the reduced function of the muscles involved. Fatty re-placement of the maxillolabialis muscle reflects the reduced mobility of the upper lip. The fat of the lateral aspect of the lower jaw is derived from the masseter muscle, the function of which has been almost totally taken over by the powerful temporalis muscle. The jugal or zygomatic arch, to which the masseter is normally attached in terrestrial mammals, is reduced in odontocete cetaceans to a delicate rod of bone which is almost invariably lost during careless preparation of the skull.

The fat on the mesial aspect of the lower jaw is derived from gradual fatty replacement of the external pterygoid muscle and is correlated with reduced antero-posterior mobility of the jaw. This reduced mobility of the various muscles referred to above is unique to cetaceans. Isovaleric acid is derived from catabolized leucin of which there are large amounts in protein. Isovaleric acid is also found in substantial quantities in the excrement and urine of terrestrial mammals where fatty degeneration of the liver is involved. Varanasi et al. (1975) have demonstrated that in cetaceans isovaleric acid is synthesized into its triglycerides at an early stage in the catabolism of leucin and in this condition it is non-toxic.

If these triglycerides have an acoustic function, this is fortuitous. The fat of the blubber is synthesized directly from carbohydrate and is the homo-logue of the panniculus adiposus of terrestrial mammals.

The same authors have shown that the velocity of sound in these lipids is considerably lower than that in sea water, so that a great deal of refraction of sound waves must occur at the boundary of the two media. It should be recalled, however, that the degree of refraction of sound between two media depends solely on the ratio of the velocities of sound in the two media over a very wide range of frequencies, so that all frequencies emitted in the sonar pulse would be refracted by the same amount and would be brought to a focus at some point in front of the head. This would be a distinct disadvantage in respect of long-range echolocation.

The degree of refraction is given by the expression:

$$\frac{\text{Angle of incidence}}{\text{Angle of refraction}} \quad \frac{\sin \theta_1}{\sin \theta_2} = \frac{c_1}{c_2}, \quad \begin{array}{l}\text{Triglycerides of isovaleric acid} \\ \text{Blubber and/or sea water.}\end{array}$$

The velocity of sound in isovaleric acid (C_1) given by Varanasi *et al.* (*loc cit*) is 1367 m/s and that of sea water C_2 (1504 m/s). The value $\sin \theta_1$ depends upon the shape of the melon. It has been suggested that the shape of the melon can be altered at will but it is obvious from the anatomical point of view that this is not possible in the majority of odontocetes. It has also been postulated that the lipid constitution of the melon can be altered more or less at will. This would require a degree of conscious control over the chemical constitution of the body hitherto unknown amongst mammals.

There is little doubt that the melon helps to throw the voice forward but a fine degree of focussing capability seems highly improbable.

It has already been stated that the melon constitutes the upper lip in cetaceans—and therefore it may be concluded that it is a sensory area. It may be interesting in this respect to reiterate the remarks of Ernst Huber (1934) on this subject:

"Some authors have ascribed to the melon a possible function as a kind of shock absorber. I believe, however, that it has a far more important function, Judging by its position, its construction and its large sensory nerve supply, I suspect that it might enable the animal to appreciate changes in water pressure. It must be remembered that the sinus hairs in the region of the snout which play such an important role in the pinnipeds, the next best swimmers among marine mammals, have disappeared in the porpoise as well as other whales. Only traces of these sinus hairs are found as rudimentary structures in foetal stages. They may persist almost until the time of birth as my full-term porpoise foetus shows.

These important structures could hardly have disappeared unless another mechanism had taken over their vital function. This new mechanism I assume to be located in the melon. Kept under tension by the surrounding muscles and elastic tissue, the melon yields to external pressure. This could easily be demonstrated in the live porpoise. Any change in water pressure when the animal accelerates, when it dives to greater depths, or approaches unyielding objects, will first exert its effect on the head. Even the slightest change in water pressure may thus be appreciated by the animal, with the aid of the sensory nerves which richly supply the melon."

The whalebone whales (Mysticeti) have no melon, nor any similar structure. Through the investigations of Japha (1907) and others, it has been shown, however, that members of this group have as many as 130 peculiarly specialized sinus hairs in the snout region. These represent modified mystacial and mental vibrissae. While the bristles of these sinus hairs are vestigial, their hair sheaths persist, and the blood sinuses therein have reached further extension and elaboration. According to Japha (*loc. cit.*) every one of these peculiarly modified sinus hairs is supplied with several hundred myelinated nerve fibres from the infraorbital and mental nerves. These nerve fibres connect with numerous tactile Pacinian corpuscles which are lodged within the hair sheath and in its immediate vicinity. The occurrence of tactile corpuscles in connection with hairs is unique and confined to whales.

The extraordinarily rich, sensory supply of the modified mustacial and mental vibrissae in connection with the numerous specialized nerve endings (tactile corpuscles) would, it seems, make the vibrissae apparatus of the snout of the whalebone whales a most efficient mechanism for the appreciation of slight changes in water pressure. This mechanism could hardly have developed alone in connection with food pursuit, as Japha suggested, but most probably serves the purpose of a hydrostatic organ, taking over the supposed function of the melon in the Odontoceti. Naturally, such a delicate sensory apparatus at the same time might serve as a valuable aid in a search for food.

Thus while the toothed whales, with the exception of *Platanista*, are probably guided chiefly by eye-sight in their pursuit of food, the whalebone whales presumably rely also on their oral sensory apparatus. It is thus obvious that both the toothed whales and the whalebone whales in pelagic life are dependent on a highly efficient hydrostatic organ."

We heartily concur with Huber's views and may add that the melon probably also acts as an organ of dynamic orientation. It is well known that the semicircular canals are considerably reduced in size in relation to the size of the auditory labyrinth, remarkably so in the Mysticeti. The melon would be extremely useful in mid-water conditions where there are no visual references and the light values are more or less uniform, for we consider that the melon would also be sensitive to rotational, lateral and ventral movements of the head and consequently of the whole body. Figures 34 and 35 show the distribution of branches of the trigeminal nerve and the arterial system of the melon respectively. The venous system is even more complex and cannot be depicted. The venous channels occupy the interstices between the adipose cells and are without a well-defined endothelium.

The melon appears non-vascular when dissected in dead specimens and this must be attributed to rapid post-mortem vaso-constriction as microscopic examination reveals numerous erythrocytes in the vascular systems. Injection with radio-opaque materials renders the whole melon practically opaque to X-rays. From the above description it may be inferred that the melon is not an amorphous mass of fat, is highly unsuitable as an "acoustic lens" and can hardly be referred to as "acoustic tissue".

VI. Role of the Upper Nasal Passages in Phonation

In order that the larynx may function in the manner described in the preceding chapter, air must flow through the glottis into the upper bony nares. Moreover, since the animal is required to use its system almost continuously whilst submerged at considerable depths and for long periods, this air must be conserved and recycled for further use. Although it has not yet been proved, it is our belief, from a study of the bronchiolar sphincter muscles in the smaller cetaceans, that prior to a period of moderately deep, or lengthy submergence, the air is actively pumped out of the alveoli of the lungs into the upper bronchial passages and trachea. The space left by the collapsed lungs is then filled by injection of oxygenated blood into the thoracic retia mirabilia, a serpentine mass of arteries and veins which occupies the whole of the dorsal wall of the thorax and the spaces between the transverse processes of the vertebrae and the upper part of the ribs. This mass of blood vessels forms the origin of the main supply of blood to the brain in cetaceans, the internal carotid being degenerate. Other functions have been assigned to the retia mirabilia by a number of authors but as they are not relevant to the subject of phonation they will not be discussed here.

A hitherto unpublished anatomical observation is that all the pulmonary, alveolar tissue of the lungs in cetaceans is confined to the external, superficial areas, the greater bulk being made up of yellow elastic tissue. This fact could be another explanation for the almost complete evacuation of the lungs between respirations.

Scholander (1940) has shown that the muscles of propulsion act anaerobically during diving, thus accumulating lactic acid. This is converted to CO_2 and water when normal respiration is restored. It has been stated that bronchiolar sphincter muscles do not occur in the larger balaenopterid whales nor in the Bottle-nosed and Sperm-whales, *Hyperoodon* and *Physeter*. However as Scholander has suggested, these larger whales probably pass straight through the danger zone to depths at which alveolar collapse occurs through hydrostatic pressure alone, thus avoiding nitrogen absorption.

Whatever the method of alveolar collapse, it is likely that during prolonged submergence, the thorax is immobilized and takes no part in phonation in cetaceans as it does in terrestrial mammals. Ridgway *et al.* (1979) found no intracostal nor intertracheal rise in pressure during audio-frequency phonation whilst an experimental animal was confined at the surface.

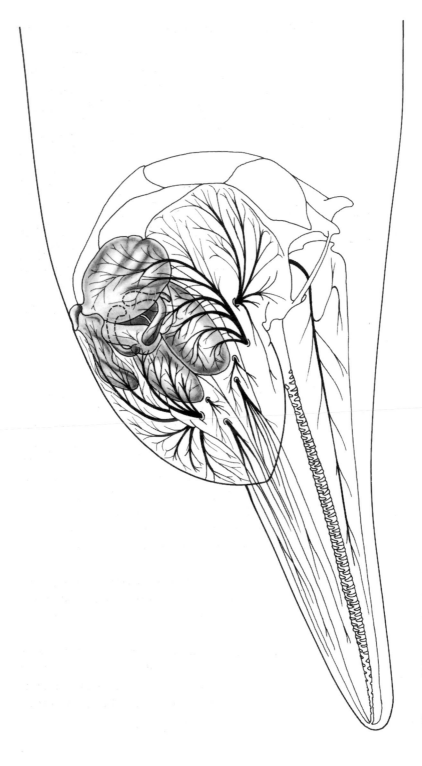

Fig. 34 Diagram showing distribution of branches of the trigeminal nerve to the melon, upper respiratory tract and rostrum of a typical delphinid.

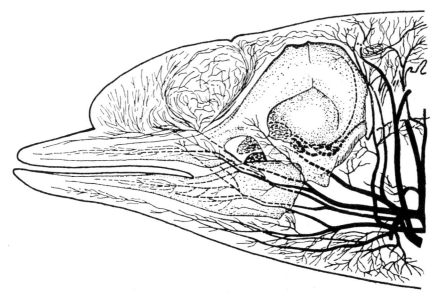

Fig. 35 Diagram showing the distribution of arteries to the melon of *Tursiops truncatus* (after Agarkov *et al.*, 1979).

For this reason alone they completely "ruled out the larynx as a primary source of sound". Had they measured the interlaryngeal pressure the results might have been quite different. We consider that all pressure changes during phonation are initiated in the larynx and cannot be transmitted posteriorly. For this reason we would not expect an intracostal nor inter-tracheal rise in pressure. It is possible that cetaceans have to "make do" in phonation with the amount of air that can be contained in laryngeal air chambers. It can be demonstrated that the large laryngeal air-sac of the balaenopterid whales completely fills the lumen of the trachea when inflated and is structurally homologous with the laryngeal air-sacs of the odontocetes.

Although the amount of air used in a burst of sonar activity is probably not known it is possible to estimate in broad terms the amount that would be used if phonation occurred by the method described in the preceding chapter.

Experimentally, a resected larynx of *Tursiops truncatus* used approximately 120 cm³ of air in the artifical production of a train of relaxation oscillations lasting 10 s with an average pressure of 25 mm Hg. This is probably an over-estimation of the amount used in life and it is interesting that the theoretical estimate is about half this quantity.

Figure 28 shows a series of three pulses with the oscilloscope set at 100 cycles/s. The recurrence frequency is therefore 300 pps. As the camera exposure was 0·04 s there are 4 to 5 clicks superimposed on each pulse and the duration of a single click is approximately 200 μs. Taking an average recurrence frequency of 150 pps during a 10 s train of pulses and an average jet diameter of 0·05 mm the volume of air can be calculated if the velocity of the jet is known. The frequency content of the pulses, according to Krüger and Schmidtke (1919), indicates that the velocity is above the speed of sound but an estimate can arrive at it in part with reference to Kinetic Theory. If the air in the respiratory tract of cetaceans reaches 300°C on the absolute thermodynamic scale and the difference in air pressure within the larynx is greater than that in the supraglottal cavities by 1/3 atm at any instant, then the difference in the kinetic energy of the molecules of air inside and outside the larynx will be 48×10^4 dynes cm² approximately. This figure represents about the limit of the force that can be exerted by the thyroarytenoid muscle in *Tursiops truncatus* but a much smaller pressure difference could be assumed. The density difference, if the air were saturated with water vapour, would be $3·5 \times 10^{-4}$ g/cm³. If this pressure difference were released suddenly at the tip of the glottis then the velocity of the escaping gas would approximate to the mean velocity of agitation of molecules of the gas along the axis of the glottis.

This is equal to $(P/p)^{\frac{1}{2}}$ where P is the pressure difference $= 2/3$ of the total kinetic energy and $p = $ density difference $= [(32 \times 10^4)/(3·5 \times 10^{-4})]^{\frac{1}{2}} = 30·2 \times 10^3$ cm/s. If the air is further accelerated by elastic recoil of the glottic cartilages to slightly above the speed of sound as Husson (*loc. cit.*) states, then the amount of air used during a burst of sonar lasting 10 s with three laryngeal jets operating simultaneously with 0·2 ms pulses at 150/s

$$= 33·3 \times 10^3 \times \pi (0·025)^2 \times 150 \times 10 \times 3 \times 0·0002 = 60 \text{ cm}^3$$

It should be pointed out that the above calculation is meant to apply to an animal about the size of *Tursiops* but because pressures and densities are proportional, the jet velocity would be about the same for smaller values of P and p. The size-specific variables are considered to be intensity and recurrence frequency.

The premaxillary air-sacs in *Steno* are relatively small and contain a maximum of 50 cm³ of air when inflated. This probably represents the limit of the volume of air that can be used in a single burst of sonar. It has been postulated that the sound-producing mechanism is located in the upper nares and is probably represented by the two valvular flaps which are to be found on the lateral borders of the nasal plugs.

This idea is rejected because it has been found to be mechanically impossible for experimental and anatomical reasons. It was stated earlier that air could not be blown through the trachea of a fresh-dead specimen of a porpoise or dolphin for more than 10 s unless the blowhole were intubated. This is due to the fact that pneumatic self-sealing action of the blowhole as described by Lawrence and Schevill (1965) is so powerful that continued air injection usually results in failure of the pipeline supplying the air.

The action has been demonstrated schematically in Fig. 36 but the diagrams should be consulted in conjunction with Figs 20–24.

In Figs 36 A and B, the blowhole has been dilated by the maxillonasalis and procerus nasi. The epiglottic spout is also dilated and drawn forward into the nasopharynx. The laryngeal, and tubular air-sacs are virtually empty and the palatopharyngeal sphincter relaxed. This represents the condition during exhalation and inhalation.

In Fig. 36 C the blowhole muscles have been relaxed and the nasal plugs have moved back over the nares. The air continues to flow to the highest point in the head and inflates the tubular sacs. Meanwhile the epiglottic spout is withdrawn towards the palatopharyngeal sphincter muscle which begins to contract.

In Fig. 36 D the tubular sacs are fully inflated and the posterior sac presses down into a recess on the upper surface of the nasal plug. The anterior sac presses upward closing the aperture of the blowhole, the anterior and posterior lips of which are sculptured to make a perfect fit. At this stage no more air can enter the tubular sacs, and therefore air attempts to enter the premaxillary sacs but is prevented from doing so by transverse contraction of the functional part of the maxillolabialis muscle. In this situation the laryngeal air-sacs expand. After this there is a hypothetical situation in which the air is static and the lungs evacuated prior to a dive.

In Fig. 36 E a burst of sonar begins due to contraction of the thyroary-tenoid and thyropalatine muscles and the palatopharyngeal sphincter is strongly constricted. Air passes out through the lateral aryepiglottic folds of the glottis producing a train of sonar clicks and vibrations are carried through the palatopharyngeal muscles to the base of the rostrum. The maxillolabialis relaxes and the premaxillary sacs expand During this

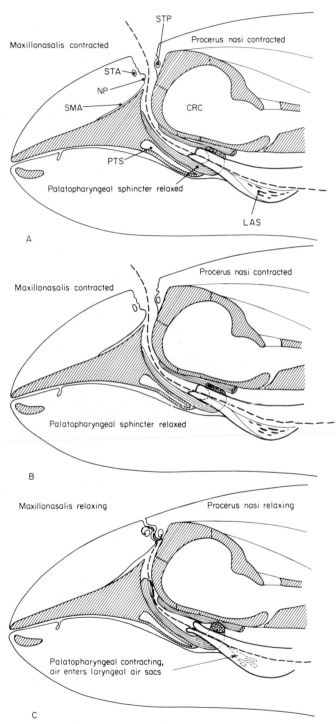

Fig. 36 Diagram showing the passage of air during respiration, followed by the occlusion of the blowhole caused by internal gaseous pressure in the tubular air-sacs and the recycling of air from the laryngeal air-sacs during sonar emissions.

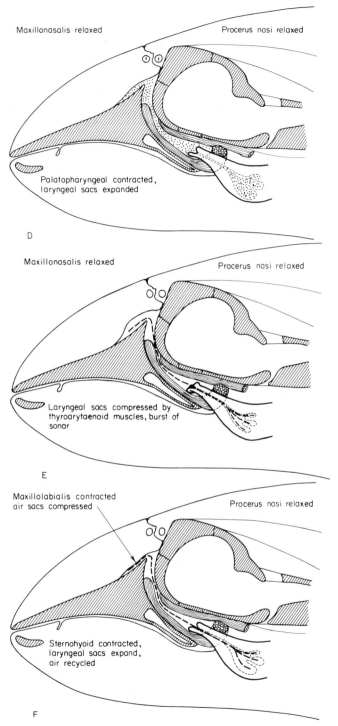

CRC = cranial cavity, LAS = laryngeal air-sac, NP = nasal plug, PTS = pterygoid sinus, SMA = supramaxillary air-sac, STA = anterior tubular air-sac, STP = posterior tubular air-sac.

operation the diagonal membranes (Fig. 24: DM) guarding the entrances to the bony nares fly against the lateral, ventral surfaces of the nasal plugs so that no more air can enter the tubular air-sacs. Moreover, no air can move from the tubular sacs to the premaxillary sacs due to the uni-directional, valvular action of the lateral flaps of the nasal plugs No air can escape from the *tubular sacs until the blowhole is dilated*

In Fig. 36 F at the end of a burst of sonar the wings of the thyroid cartilage are forced apart by the action of the sternothyroid muscle (see Fig. 8: MST), and the maxillolabialis above the premaxillary sacs contracts (Figs 22, 23: MXL); thus the air is drawn and compressed downward into the laryngeal air-sacs and made ready for another burst of sonar

By sealing the slots (SL) which lead from the upper nares into the posterior tubular sacs it is possible to inflate the tubular sacs with air through the connecting sac by means of a hypodermic needle It was found that the combined sacs took a little less than 20 cm^3 of air, and it is doubtful whether this figure could be more than doubled under hydrostatic pressure. As Lawrence and Schevill have pointed out, manual pressure on the sacs in the living dolphin causes air to be extruded upwards through the lips of the blowhole into the vestibular sacs (Fig. 21: VS) which lie dorso-lateral to the tubular sacs This air escapes through the slots (SL) which lie *above* the level of the nasal plugs

It has already been stated that it would be possible to produce whistling noises by the passage of air through the valvular extensions of the nasal plugs, but because of the relatively large amount of air required for this purpose and the small capacity of the tubular sacs, such whistling would always be accompanied by loss of air into the vestibular sacs, thence to the water, if the duration of the whistle exceeded more than a second or two.

Lawrence and Schevill have ascribed the function of a water trap to the vestibular sacs and there is no doubt that this is one of their main functions. A jet of water with a head of 2 atm aimed directly at the blow-hole of the head of *Steno* referred to above, caused the vestibular sacs to expand to their full capacity and then overflow, but no water was observed to flow from the severed end of the trachea. Since many cetaceans have been observed to exhale before reaching the surface a water trap would appear to be essential. However S. Andersen, University of Odense, in a private communication, stated that in *Phocoena* these sacs act as flotation devices when the animal is stationary or sleeping, ensuring that the blowhole remains clear of the surface when breathing. Recently dead animals have been observed to sink below the surface when the vestibular sacs have been punctured. Both of these explanations of the function of the vestibular sacs raise the question of the asymmetry of the cetacean skull and upper nasal mechanisms. It has frequently been postulated that this asymmetry is associated with the sonar system although it is difficult to imagine from the physical standpoint what advantage could be gained from such asymmetry.

Our view is that the total asymmetry of the cetacean skull and nasal system is associated with the mechanics of respiration, and the remarks of Andersen are relevant to one aspect of this function. The cetacean "image" is that of a very active animal constantly moving through the ocean at high speed. It is well known that like other mammals, whales and dolphins require rest and sleep and must seek relatively undisturbed surface conditions for this purpose. It would be of considerable disadvantage in these circumstances if the blowhole were placed centrally on the top of the head.

According to Lawrence and Schevill the average combined expiratory and inspiratory duration of the blow in *Tursiops* was 0·75 s and was frequently accompanied by a certain amount of water spray. In common with all mammals, the thoracic mechanism is such that inspiration is more rapid and positive than exhalation. If the blowhole were vertically positioned a considerable quantity of the exhaled water spray and gas would be rein-haled on the inspiratory phase. This would obviously not occur in a dolphin in motion or in windy conditions.

If, during the course of evolution of the lateral displacement of the blowhole, the entire system of pneumatic cavities had remained sym-metrical about the plane of the mesethmoid, the head would be hydro-statically unstable and there would be a constant tendency for the head to rotate until the blowhole was again top-dead-centre. It is clear that during the lateral displacement of the blowhole and nasal septum, the air-sac system of the right hand side has become "parasitic" on that of the left, so that the whole system of air spaces remains approximately on the central axis of the skull and is therefore hydrostatically stable. The extreme example of this process is found in the Sperm-whale, *Physeter catodon*, where the entire system of air-sacs on the left hand side has disappeared, whereas that of the right is greatly hypertrophied and centrally disposed about the long axis of the head.

Although the plane of the mesethmoid makes an angle to the left of the sagittal plane, the long axis of the lateral diverticula of the nares, the lips of the blowhole, and the chord of the crescent of the blowhole remain at right angles to the plane of the mesethmoid, so that the entire systems of the two sides have an antero-posterior displacement.

It follows that since the structures of the right side are larger than those on the left and are displaced antero-posteriorly, then the muscles which control these systems must also be larger and antero-posteriorly displaced. Finally the bones of the skull which form the insertions of these muscles must be commensurate with the soft parts so that the entire upper surface of the skull takes on an asymmetry that is subservient to the proper functioning of the soft parts.

Schenkkan (1972) has criticized this idea on the grounds that in *Pontoporia* only the vestibular sacs are asymmetrical, the rest of the air-sac system and skull being more or less symmetrical, but it is clear from the fossil record that throughout the evolution of the Cetacea asymmetry has taken place from the top of the nasal tract downwards and that asymmetry of the skull is only one of the more recent developments, as are the pre-maxillary sacs. Perhaps if cetaceans are allowed to continue the process of evolution, the skull will become even more asymmetrical than it is now.

In most odontocetes the left bony narial aperture is larger than the right and the naris follows a more vertical course. The epiglottic spout is bent slightly to the left (markedly so in the Ziphiidae and Physeteridae) and when the larynx is pushed upward into the nasopharynx the glottis moves towards the left posterior naris.

All this evidence suggests that in the Odontoceti respiration takes place predominantly through the left naris. This is most obvious in the Physeteridae in which the diameter of the bony aperture of the left side is five to seven times greater than that of the right, and the nostril of the left side forms a continuous tube, the calibre of which is controlled by circular muscle throughout its length.

It is possible that the whole arrangement is conducive to more efficient external respiration, and in having greater survival value is probably the result of natural selection. It can be demonstrated by a simple calculation that a pair of nostrils of equal calibre, such as is found in terrestrial mammals, is a relatively inefficient arrangement as far as ventilation of the lungs is concerned. Consider two circular pipes of diameter 4 cm and 1 cm length. Then the internal surface area is given by $2\pi R$ and the volume by πR^2 for each pipe. The ratio of total surface area to total volume for both pipes:

$$= 2\,(2\pi 2): 2(\pi 2^2)$$

the value π is a constant and can be neglected, therefore the ratio is:

$$8:8$$

Suppose now that the diameter of one pipe is reduced to 2 cm and the other increased by a corresponding amount, becoming 6 cm, the ratio of total surface area to volume becomes:

$$(2\pi 1):(2\pi 3):(\pi 1^2) + (\pi 3^2)$$

the ratio is: $8:10$

Thus at a given velocity, a greater volume of air could be passed through the two unequal apertures for the same surface area, than that for the two equally sized apertures. If passage through the smaller aperture were suppressed completely, the ratio would become:

$$(2\pi 3):(\pi 3^2)$$
$$= 6:9$$

In this case the surface to volume ratio is even more favourable and the total volume of air passed through the single, larger aperture, is again greater than that through both the original pair of 4 cm diameter. The greater one aperture becomes with respect to the other, the greater the advantage from the ventilation point of view, and it is our opinion that this principle is also involved in the asymmetry of the cetacean skull which finds its extreme expression in the Sperm-whale. It is clear that in Physeteroidea the smaller aperture plays no part in external respiration.

In commencing this discussion we stated that the arrangement of paired nostrils of equal diameter was an inefficient mechanism from the point of view of breathing.

Why then is it present in all terrestrial vertebrates? It is our opinion that the explanation is to be found within the cognizance of the whole purpose of bilateral symmetry in living organisms. Efficient motion in a chosen direction became possible in evolutionary history, only when sensory and locomotory organs of various kinds became symmetrically disposed about a central axis of the body. By such an arrangement animals became able to detect differences between the environment on the two sides of the body and to move towards a preferred environment without experimentation.

It is this general principle which we believe to be embodied in the possession by all active vertebrates of paired nostrils, and in the forked tongue in certain reptiles. The nostrils are the seat of the olfactory conchae and are, therefore, primarily sensory organs. Just as it is necessary to have two ears for directional hearing, so it is necessary to have two nostrils for the purpose of directional olfaction. Modern man has come to be almost totally independent of the directional properties of his sense organs and tends to neglect the importance of this aspect of their physiology when studying the lower vertebrates. At first sight it may seem that the nostrils are not sufficiently widely separated to form a time or intensity base for directional olfaction. However, the planes of the nostrils in most animals lie across widely divergent axes, and air is inhaled and exhaled from a considerable distance along these axes, as may be deduced from simple laboratory experiments and by the observation of the columns of water vapour expelled from the nostrils of animals in cold temperatures. Most animals are able to follow with great accuracy an invisible, undulating scent trail on the ground. As an example, if the leg of a frog be trailed in a winding course across a piece of white card and the card be allowed to dry, there remains no visible trace of the track of the foot. Nevertheless, a grass snake can follow this trail with perfect accuracy. It is possible that this is achieved through the detection by the organs of Jacobson of slight differences in the scent on the two prongs of the forked tongue. If a dog were forced to experiment by quartering the ground every time a scent changed direction, a great deal of time and muscular effort would be required—in other words, the system would be inefficient. It is conceivable that when an animal "sniffs the wind" away from the ground, the nostrils are used in conjunction with the extremely sensitive, tactile vibrissae, which lie in regularly spaced rows on either side of the nose and probably act as wind direction indicators. The idea that these are used solely for measuring the distance apart of obstacles is, in our opinion, erroneous.

Olfactory organs are absent in odontocete cetaceans and the blowhole is entirely subservient to breathing; thus there is no need for paired nostrils. The blowhole is, therefore, adapted for more efficient ventilation. For two examples in the Vertebrates which "prove the rule" one may draw attention to the organization of the olfactory organs in a group of Devonian fishes, the cephalaspids, which led a completely sedentary life on the bottom of the sea. In these fish there is but one nostril, mesially placed, but there is clear evidence internally that at some time in their evolutionary history they possessed two external nostrils.

The other example is in the paired nostrils of the Mysticeti, the baleen whales (Fig. 37), which possess both olfactory conchae and olfactory lobes that the latter two structures are relatively small, and that the animals live permanently in the sea has led to the supposition, by some cetologists, that the sense of smell is either reduced or absent in the Mysticeti also. When it is remembered that the Mysticeti are plankton feeders, and that plankton-rich sea water frequently has a distinctive odour, it is not difficult to imagine that these large animals may "sniff the wind" in search of their food. It is doubtful, however, whether there is any intra-narial directionality in the olfactory sense and it is possible that the tactile vibrissae on the surface of the head can act as long-range wind direction perceptors as well as underwater close-range receptors (Purves, 1966).

Perhaps the most striking evidence in favour of the hypothesis of a laryngeal source of phonation in cetaceans is that afforded by the Mysticeti. Schevill et al. (1964), who first established that the persistent 20 Hz sounds recorded in certain parts of the Atlantic were those emitted by the Fin-whale, Balaenoptera physalus, has compiled a list of those mysticetes known to emit sounds, together with the authorship of the observations. Perkins (1966) published sonograms demonstrating that Fin-whales were also capable of emitting whistles and chirps with frequencies up to 5 kHz. Schevill et al. (1964) drew attention to the laryngeal sac in the Mysticeti and indicated its possible use in phonation. Since most of the published work on the larynx of mysticetes concerns foetal and juvenile specimens it might be worthwhile drawing attention to a few details concerning the subadult larynx which do not seem to have been described. In 1955, Purves had examined the larynges of 18 Fin-whales at Steinshamm, Norway, and two of these were taken to the British Museum (Natural History) for dissection and exhibition. A diagram of the larynx of Balaenoptera physalus is given in Fig. 38.

In all the specimens examined there was a semidiscoid flange projecting forward from the apex of each arytenoid cartilage. The flange was about 8 cm in radius and about 1 cm thick. In the more adult specimens the mesial surface was covered with closely-packed, warty rugosities (Fig. 38: insert) of approximately 1 mm diameter so that the whole surface presented an appearance similar to that of the tongue in terrestrial mammals. Microscopic examination revealed a thick, cornified epithelium with a very

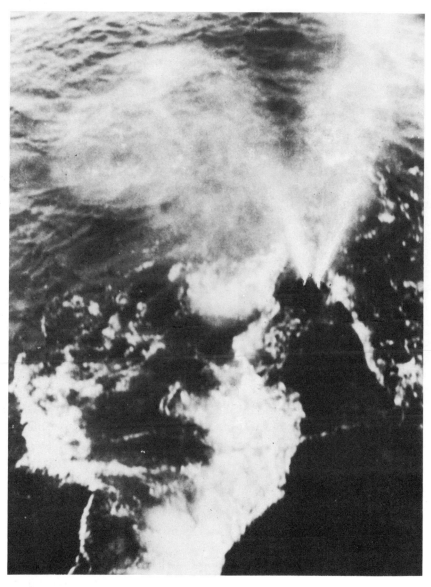

Fig. 37 Photograph showing the double "blow" of a black Right whale, *Eubalaena australis*.

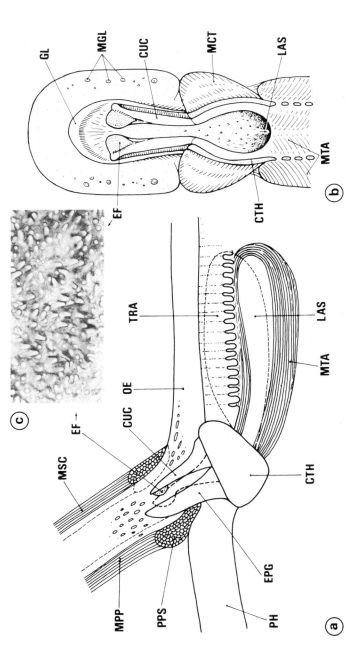

Fig. 38 (a–c) Diagram showing the structure of the larynx in *Balaenoptera physalus* (a) lateral view showing laryngeal air-sac and epiglottic spout in open, and closed (dotted line) position. (b) larynx cut open from the rear to show entrance to laryngeal air-sac (c) cornified rugosities on mesial surface of epiglottic flange.

CTH = thyroid cartilage, CUC = cuneiform cartilage, EF = epiglottic flange, EPG = epiglottis, GL = glottis, LAS = laryngeal air-sac, MGL = mucous glands, MPP = palatopharyngeal muscle, MSC = superior constrictor muscle, MTA = thyroarytenoid muscle, OE = oesophagus, PH = pharynx, PPS = palatopharyngeal sphincter, TRA = trachea.

deeply papillated zona germinative like that of the external skin, and it is clear that these surfaces were subject to considerable wear. When the arytenoid and epiglottic cartilages were approximated and laterally compressed as if by the arcus palatinus, the mesial surface of each flange became closely opposed and the antero-ventral borders of the double assembly fitted deeply into the trough formed by the epiglottis. The flanges themselves were composed entirely of elastic tissue and it is certain that if air were blown through the glottis from below with sufficient pressure it would cause the system to vibrate harmonically and probably with a fixed frequency, whether the larynx was used for phonation or not. In the more juvenile specimens, the flanges were smaller and their mesial surfaces smoother although incipient rugosities were present. It is possible that the formation of the rugosities was with the onset of sexual maturity. The origin of these arytenoid flanges is not difficult to find when the foetal specimen is examined. The "arytenoid body" as described by Carte and Macalister (1868) and by Schulte (1916) is wholly semidiscoid in shape and consists of a very small posterior arytenoid cartilage and a large anteriorly directed membrane which is homologous with the phonatory "arytenoid fold" of the ungulate mammals. During the post-natal lengthening of the arytenoid cartilage and the epiglottis, this membrane is carried dorsally so that it becomes remote from the main body of the larynx. Its functionality in the adult is indicated by its isogonic growth in diameter and thickness, and its situation at the apex of the glottis gives support to the notion that the phonatory region of the larynx is, as in the Odontoceti, at the tip of the epiglottic spout.

In a female 17·75 m long the laryngeal pouch was 120 cm in length, 40 cm in depth and 30 cm wide. When the animal was lying on its back the whole sac fitted neatly into the lumen of the trachea, in this part of which the cartilagious rings were fused into a solid plate dorsally and absent ventrally. The great mass of the thyroarytenoid muscle, which formed the ventral and lateral walls of the sac, was 20 cm thick at its greatest width. The muscles of this system have been adequately described by Hosokawa (1950) and it is apparent that the sac and its muscle system is homologous with the laryngeal air-sacs and their closing muscles in the Odontoceti. It is clear from the unique construction of the trachea and the muscular arrangement that the sac can be drawn up into the lumen of the trachea so that it isolates the supraglottal air spaces from the thorax. It can be conceived that an arrangement such as this could be used to evacuate the lungs if this did not occur through hydrostatic pressure at moderate depths. Since there are apparently no distensible air-sacs in the superior nares, phonation by the larynx would require more pressure than in the Odontoceti, but the muscles of the laryngeal sac seem quite adequate to perform this function and to produce the large energy outputs which have been measured.

In the Mysticeti, as in the Odontoceti, the larynx is coupled anatomically to the bones of the rostrum by the palatopharyngeal muscle complex and it is conceivable that it is so coupled acoustically. It is significant that in common with smaller cetaceans, the component bones of the two sides of the rostrum remain separated from each other by the vomer throughout life. The length of a 20 Hz sound wave in water is 70 m approximately, therefore very little directionality could be imparted to the 20 Hz pulse by the "double-source" arrangement, as described for the Odontoceti. However, the wavelength of the 5 kHz "chirp" described by Perkins (1966) is only 30 cm. The width of the skull in an adult Fin-whale can be in excess of five times this figure, therefore the "chirp" must have very definite directional properties, if propagated by the skull.

Such a laryngeal sac is also present in the Humpback-whale, *Megaptera novaeangliae*, and could be responsible for the now familiar "song of the Humpback-whale".

Since there are no nasal plugs in the baleen whales and the laryngeal sac is obviously of phonatory significance it will have to be conceded (in the absence of any published information about an alternative source of sound) that at least in these large whales the larynx *is* the source of sound. But what of the great toothed whale, the Sperm-whale? There are no *recognizable* nasal plugs in this animal either, so not surprisingly an entirely new structure has been identified as being the source. This is the entrance to the right nostril which the French anatomist Beauregard (1894) referred to as the "museau de singe" or monkey's mouth because of its superficial resemblance to that part of the monkey's physiognomy. It would seem that to some cetologists any anatomical structure can be given the credit for producing the sound except the true vocal organ, the larynx. Most of these ideas could be ignored, had they not been so widely publicized in the popular and semi-scientific press.

According to Clarke (1979) asdic records provide good evidence that Sperm-whales not infrequently dive to in excess of 1000 m off Durban, whereas Heezen (1957) states that they appear to be able to dive to 1500 m.

It is difficult to see how the whale can make any noise at all at these depths yet echolocate it must, since there is almost total darkness in these regions of the sea.

During the course of its evolution the position of the blowhole in the Sperm-whale has moved twice. First to a position on top of the head as in other whales and dolphins and secondly forward again to the tip of the snout. This has been due to the emergence of an entirely new structure, the spermaceti organ (Fig. 39: SO) which has grown like a sebaceous cyst in the posterior wall of the right nostril under the mucous membrane of the cartilage of the nasal septum. The spermaceti organ has become enormously enlarged and with it the whole head, so as to occupy about a third of the animal's length. The nostrils too have become greatly length-ened. If the growth of the spermaceti organ is observed from early foetal

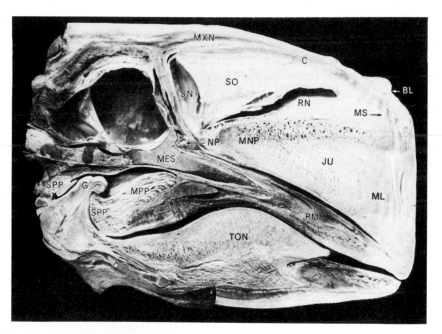

Fig. 39 Sagittal bisection of a 150 cm foetus of a Sperm-whale, *Physeter macrocephalus*.

BL = blowhole, C = "case", G = glottis, JU = junk, MES = mesethmoid, ML = melon, MNP = nasal plug muscle, MPP = palatopharyngeus muscle, MS = "museau de singe", MXN = maxillonasalis muscle, NP = nasal plug, PM = premaxillary bone, RN = right naris, SO = spermaceti organ, SPP = palatopharyngeal sphincter, TON = tongue.

specimens it can be seen to push the cartilage of the nasal septum to the left hand side. The right nostril (RN) has become greatly enlarged and partially wrapped round the spermaceti organ like a deflated balloon. The left nostril has meanwhile remained relatively small in calibre.

Paradoxically, at the skull end, the bony naris of the right nostril is about five times smaller than the left (Fig. 40). For various anatomical reasons we believe that normal respiration takes place entirely through the left nostril. During diving we believe that all the exhaled air from the lungs is passed into the capacious right nostril and retained there for phonatory purposes. According to Clarke (1970) the lung capacity of a Sperm-whale is about 1000 litres but it is obvious that the capacity of the left nostril and nasofrontal sac in an adult Sperm-whale when inflated would be considerably more than this, so it would seem that more than one exhalation must take place before the right nostril is fully inflated. It may be of some significance that in most of the illustrations of open-boat Sperm-whale fishing the upper part of the head is shown well clear of the water prior to a dive. This may be due to increased buoyancy of the head through inflation of the right nostril.

At 1000 m depth the pressure is in excess of 100 atm so that the gas in the right nostril would be reduced in accordance with Boyle's Law to slightly over 0·11 of its initial volume. However, this would be enough to allow the larynx to function in the manner described by us for the production of echolocating pulses.

The oil in the spermaceti organ solidifies to become a crystalline wax at temperatures below 29°C. Clarke (1979) has suggested that this phenomenon is made use of in the Sperm-whale to adjust the density of the head to that of sea water to maintain buoyancy at depth. The cooling of the spermaceti is achieved, he postulates, by the circulation of sea water round the right and left nostrils. However, there is no connection between the right and left nostrils except in the region of the nasopharynx in which case the glottis would also be enveloped in sea water. The entry of sea water would require the displacement of air from the nostrils and nasopharynx, so that when the water was pumped out again there would be no air left throughout the entire respiratory tract. Any pneumatic form of sound production either by the method propounded by us or that described by Norris would then be impossible.

Clarke's is a very ingenious hypothesis but if solidification of the spermaceti does take place we suggest that it is achieved by normal heat radiation and vaso-constriction as occurs in other parts of the body. If

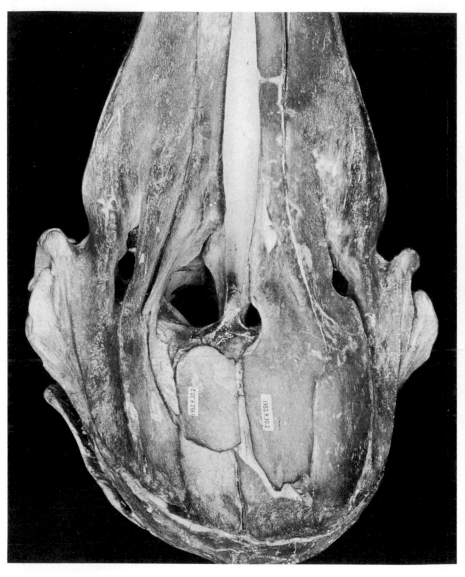

Fig. 40 Dorsal view of the skull of a foetal Sperm-whale, *Physeter macrocephalus*, to show asymmetry of the nares.

vaso-dilation is involved then heat would be lost throughout the entire body not just from the spermaceti organ. The blood circulation of the head is too complex to be described here but we regard the entry of cold sea water into the nostrils as being highly improbable, not only because of bodily heat loss, and loss of phonation, but also because of the entry of foreign particles, sea organisms, ectoparasites, etc. The production of sonar pulses in the Sperm-whale has been well documented by Watkins (1977) and Watkins and Schevill (1975).

If 80% of the air of the lungs is expelled during respiration and 90% from the respiratory tract during cooling of the spermaceti organ, what is the origin of the blow which is so forceful and prolonged when the Sperm-whale surfaces after a deep dive?

With the large pelagic species the behaviourist has to rely entirely on sound recordings made at sea. The experiments that have been carried out almost world-wide are admirable in themselves but it seems to the authors that the interpretation of the results leaves much to be desired.

From the anatomical studies two facts have emerged. All cetaceans, whatever their phylogenetic origin, have a well-defined larynx which is built on the same general plan but which is far too complicated in structure to account for the mere process of ventilation of the lungs. Strangely enough, those delphinids which have been most frequently experimented upon have the most complicated larynges of all, and yet according to a substantial number of zoologists in the United States, the larynges of these small cetaceans have completely lost their phonatory function. The possession of a complicated larynx in these cetaceans would seem, therefore, superfluous. In almost every case it is possible to predict from the structure of the larynx whether or not a cetacean will be able to produce whistles as well as sonar pulses.

IV.   Dolphin Sounds (*Sousa plumbea*)

For the accurate determination and preparation of polar diagrams of the sound energy distribution of echolocating pulses of dolphins, measurements must of necessity be made in close proximity with the animals whilst they are confined to a small area as in an aquarium. Such measurements are commonly referred to as near-field observations but give no indication of the range of the sonar field in the natural environment.

Pilleri *et al.* (1976) whilst carrying out ecological and population studies on the blind Indus dolphin *Platanista indi* during the years 1971, 1973 and 1974 were able to record sounds made by the Plumbeous or lead-coloured dolphin *Sousa plumbea* near the Kudi and Piti mouths of the Indus delta. All the observations took place between November and February.

Most species of cetaceans have, in addition to the high frequency sonar clicks, an extensive repertoire of other sounds in the low frequency range. The following is a description of the most important behaviour patterns. *Sousa* swims in a leisurely fashion. The adults are rather plump and have a long powerful dorsal fin (Fig. 41). They usually swam slowly either in pairs or in small groups of 3 to 12 individuals of different sizes. Often a single dolphin was seen or a female alone with its baby, but no large schools. The young dolphins were often seen playing and would leap into the air with their whole bodies out of water, a behaviour less common among adults. If they were followed in the boat, the dolphins would increase their normal swimming speed coming up two or three times in front of the bows to breathe, but keeping a distance of from 5 to 10 m ahead of the boat. Afterwards they would dive deeply, change direction under water and re-appear a long way behind the boat, an escape reaction common in cetaceans.

Dolphins approaching the boat could often be heard before they were seen. Click sounds were especially audible at a distance of 200–300 m when

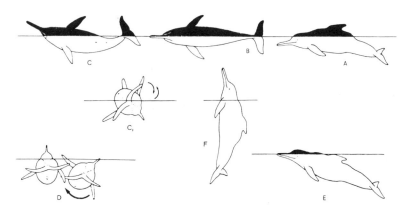

Fig. 41 Special behavioural patterns of *Sousa plumbea* in Kudi Creek, Indus delta, January 1974. A–C = rolling over and showing one flipper above the water, then rhythmic beating on the water with the flipper (C_1), D = swimming in pairs (♂♀?) and a dolphin rolling over (♂?) beside its companion, E = vertical "standing" with head out of water, F = sleep position (semi-schematic).

the dolphins were swimming directly towards the boat, but as soon as they came closer and either turned aside or swam underneath the boat, the clicks suddenly became quieter, whereas whistles and screams remained distinctly audible, although the latter were not audible at more than 100 m. This observation points to a strongly directional emission field for the sonar clicks.

In February 1974 a special kind of behaviour was observed in single or paired animals. The single dolphin or one of a pair would suddenly brake and glide forward slowly, turning on to its right side and then remaining motionless for 3 to 4 seconds in a slightly curved position with the beak and melon out of water. The dolphin would then stick its left flipper into the air and beat it five or six times rhythmically on the surface of the water. These stereotyped movements were repeatedly seen in adult animals and when they were in pairs the other dolphin usually swam close by. Water-slapping with the flippers is a common sight to visitors to dolphinaria in which the animals are trained to do this at the end of a performance but as far as is known it has not been recorded in the wild, although it is well known to occur in Humpback whales. It was not possible to make a clear analysis of the sounds emitted during this behaviour since other dolphins were playing round the boat but such sounds as were recorded were obviously made by several dolphins. During a visit to the same area some weeks later, *Sousa* were again found in large numbers but they no longer showed traces of the behaviour just described. It was therefore concluded that the latter must have been some kind of mating display. Among other pairs observed at the same time in the same area, one of the dolphins would be seen rolling over beside its partner, which would continue to swim upright (Fig. 41, D).

Among *Sousa* both in Clarence Strait and Kudi Creek a single dolphin was observed swimming rapidly round in circles whilst other dolphins remained at a distance. In the same area other pairs of dolphins were bobbing up vertically with the whole head out of the water as far as the flippers, both animals remaining belly to belly in an attitude recalling the position known to be adopted by other dolphins during mating. Other individuals were to be seen "standing" in the water, remaining motionless for a few seconds with the head above water as if inspecting the world about them (Fig. 41, F). Motionless "lying" in the water for a few seconds was observed, the dolphins remaining in the normal swimming position with only part of the melon and the back showing above water. This behaviour was interpreted as a short sleeping phase (Fig. 41, E).

During the observations there were trawlers fishing for prawns using dragnets in the mouths of the creeks. The noise of the propellers of the 15 m boats could be heard for hundreds of metres underwater and often interfered with the recordings of the dolphin sounds since the frequencies reached as high as 20 kHz. The *Sousa* were usually to be found near the trawlers where fish often fell into the water when the nets were hauled in. The crescendo of the propellers did not seem to bother them nor did it affect their sonar location. The sounds of dolphins swimming at a distance were often covered by the snapping of crabs and the croaking of fish (Sciaenidae, sp.). This very common fish which emits low frequency sounds is locally known as "boro". No boros were to be heard in many areas of the delta, especially in the deep water, but in the shallow water close to the shore at Piti Mouth the sounds of huge shoals of these fish were recorded.

During the first excursion to the delta at the beginning of January, groups of six to ten dolphins were found a long way from the fishing boats in places where no boro sounds were to be heard. At the time of the second excursion on 23 February, however, the picture was entirely different. Most of the dolphins were swimming alone, obviously hunting for fish and almost always in "boro" areas, often close to huge shoals. Creaking sounds were much less common, whereas the number of whistles and clicks remained unchanged.

I. Methodology

The dolphin sounds were recorded with a Nagra-IV L tape recorder in conjunction with a 2626 conditioning amplifier and a Bruel and Kjaer 8100 measuring hydrophone. The frequency response was limited by the tape recorder to the 30 Hz–35 kHz (—3 dB) range at 38 cm/s.

For the analyses, the dolphin sounds were transferred in the laboratory to a IV-SJ measuring Nagra. The resulting tapes were used to cut tape loops for further processing. For signal display in the time range we used a Tektronix storage oscilloscope and a Bruel and Kjaer level recorder. The sound spectra were plotted according to two different methods:

1. Sonagraphic frequency analysis using a 6061 B Kay-electric spectrograph.
2. Analysis in the form of 1/3 octave spectra using a combination of a 2607 measuring amplifier, a 1614 1/3 octave filter and a Bruel and Kjaer 2305 level recorder. For more detailed frequency analysis of a number of signals we used a specially designed narrow band filter (selective filter, Q adjustable between 1 and 100).

The sonagrams were plotted using contour display (Fig. 42).

The contour sonagram displays the results of analysis by showing equal sound levels in the same shade of black and surrounded by a contour line; the successive shades of black each represent a difference in sound level of 6 db. The fact that there is a total of seven different shades means that, unlike normal sonagrams, the dynamic range is 42 dB (Müller and Oelberg, 1976). This type of sonagram is comparatively easy to read and the intensities of the different time or frequency ranges can be compared with each other. A linear frequency axis was chosen for the display.

Click sounds are not shown by means of sonagrams since, in the light of our previous experience, we have found that this method is not particularly suitable for short signals (Pilleri *et al*, 1976a). A discussion of additional methodological problems in connection with frequency analysis of very short signals is to be found in Pilleri *et al.* (1976a and b).

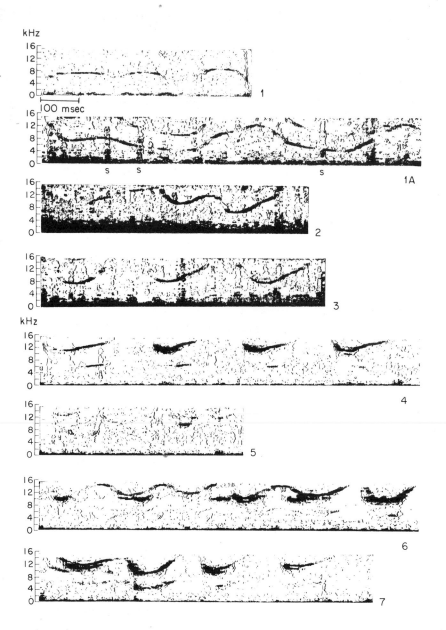

Fig. 42 Sonagram of a series of 15 screams. The sounds of several dolphins are superimposed on screams. Filter bandwidth 600 Hz. 1: sonagram of two whistles of *Sousa plumbea*. 1A: sonagram of a series of screams. In addition to the screams a few sonar clicks can be indistinctly recognized (s). For the analysis the sounds were two times frequency transformed.

II. Definitions of Types of Sounds Observed

1. **Clicks:** Signals which can be broken up into a series of single pulses. The single pulses were repeated at frequencies ranging from a few Hertz to about 500 Hz and had an average length of 100–150 μs, with 1·5–3·5 periods per pulse.

The pulses had wide-band frequency spectra with the main energy component at 20–25 kHz (cf. bandwidth of recording apparatus). Probable significance: function as sonar sounds.

2. **Whistles:** Sinusoidal, usually frequency-modulated sounds of very variable length (ms to s), occurring either singly or in series. Frequency range 3–20 kHz. Probably used for communication system.

3. **Screams:** Unpulsed sounds with a harmonic structure. Usually frequency-modulated. Always occurred in series. Frequency range 3–30 kHz. Change to whistle signals frequent. Probably used in communication system.

A. Clicks

The individual signals can be described as follows: they had an average length of 100–150 μs. The number of periods was between 1·5 and 3. The dominant frequency was at 10–30 kHz. In many cases there was also a subdominant at 8–12 kHz. The repetition rate varied between about 10 Hz and 500 Hz, with an average at 10–15 Hz.

When listening to the tapes our ears could perceive differences which led us to divide all the signals heard into two groups:

1. A series of clearly distinguishable single pulses, changing to a crackling sound at a high repetition rate. This sound was extremely common (Fig. 43).
2. A noise sounding like the creaking of a rusty door hinge. This sound was comparatively infrequent. It was often heard in combination with screams. In the oscilloscope creaking could also be broken up into a series of individual clicks (Fig. 44).

Objectively the two groups were hard to distinguish since variability was relatively high and individual features of both groups might overlap.

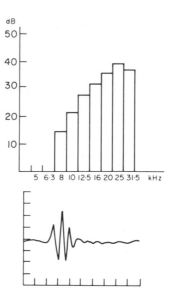

Fig. 43 Clicks from signal group 1. Oscillogram and frequency analysis. Below 8 kHz, signal in noise. Peak-to-peak. Oscillogram 50 μs/div.

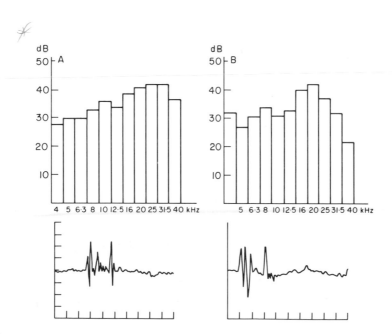

Fig. 44 Clicks from signal group 2. Frequency analyses of two clicks from different click trains (A and B). Peak-to-peak. Oscillogram 50 μs/div.

Thus the first group had repetition rates of <5 Hz to 350 Hz ($\bar{x} = 23.6$, $n = 145$, $s = 33.4$). In the second group the repetition rates were 40 Hz to 520 Hz ($\bar{x} = 268$, $n = 5$, $s = 185.8$). The two ranges almost overlap. The values in group 1 are concentrated at between 10 and 25 Hz, while those of group 2 have a wider dispersion (Fig. 45).

The length of the individual signal was very similar in both cases. Only in one case in group 2 was the click duration ten times longer than normal. On the other hand the fluctuation width for the number of periods in a click was different in the two groups: group 1: 1.5–2.5 periods, group 2: 2–3.5 periods. In both groups the dominant frequency was between 15 and 30 kHz; in most cases it was in fact between 20 and 25 kHz ($\bar{x} = 20.9$, $n = 33$, $s = 4.49$).

The falling off in the dominant frequency towards the upper frequency limit set by the tape recorder was, where it occurred at all, relatively slight. It is not impossible that this falling off in the upper spectrum range

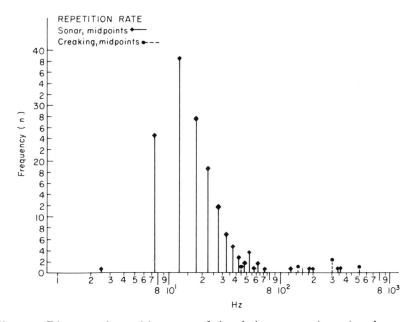

Fig. 45 Diagram of repetition rates of signals in group 1 (sonar) and group 2 (creaks). The diagram shows the midpoints for each class (class interval = 5 Hz). The abscissa indicates the repetition rates. For reasons of clarity they are shown in logarithmic form. The ordinate indicates the number of measurements per class (frequency).

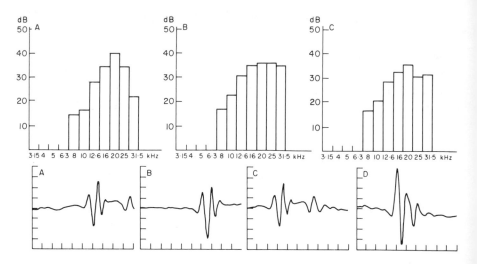

Fig. 46 Clicks from signal group 1 (sonar). Four clicks from the same click train. Frequency analyses peak-to-peak. Oscillogram 50 μs/div. Below 8 kHz, signal in ambient noise.

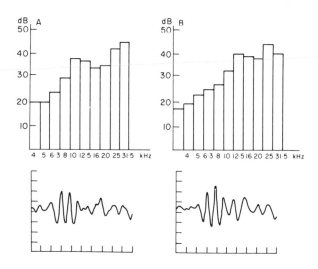

Fig. 47 Clicks from signal group 1. An indication of two-peak spectra in clicks with directly following echo. Examples of single pulses from two neighbouring click trains. Train A had a repetition rate of about 300 clicks/s, train B a rate of about 30 clicks/s. Frequency analyses peak-to-peak. Oscillogram 125 μs/div.

was caused by the recorder. Signals in group 1 had a typically simple spectrum with one uncertainty peak and one dominant frequency (Figs 43 and 46). In the case of signals with directly following echoes, there was an indication of two-peak spectra (Fig. 47). Signals from group 2 all had spectra with at least two uncertainty peaks.

In the light of the above-mentioned data it is probable that the click signals recorded can be divided into two different groups. One sound which was recorded only once and made a noise like one of the creaks, had certain characteristics which did not fall into either of the two categories: with a repetition rate of 140 Hz, the duration of the individual clicks was 1·5 ms, the number of periods 2–3 and the frequency of the most intensive period 1·7 kHz. The amplitude or sound interval of the entire click train was conspicuously small.

The sounds of group 1 were very often heard either alone or in combination with other types of sound. In groups of more than two dolphins the group 1 clicks were heard at the same time as whistles and screams.

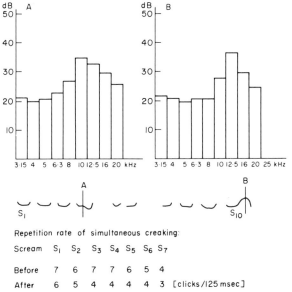

Fig. 48 Correlation between a rapid pulse train (creaks) and simultaneously emitted screams. The contours of all the sounds in the series are shown below the two frequency spectra which represent the 4th and 10th scream respectively. The table below shows the repetition rate of the emitted creaks before and after a scream. S_{1-7} = screams Nos 1–7. Numbers show clicks/125 ms.

Repetition rate of simultaneous creaking

Scream	S_1	S_2	S_3	S_4	S_5	S_6	S_7
Before	7	6	7	7	6	5	4
After	6	5	4	4	4	4	3 [clicks/125 msec]

Creaks (group 2) were much less common and heard mainly in large groups of dolphins. Creaks occurred mostly in combination with screams and occasionally alone directly after screams had been uttered. In one case the repetition rate of a creak was definitely correlated with the sound sequence in a simultaneously occurring series of screams (Fig. 48).

B. Whistles

The whistle signals recorded were either single sounds or grouped in series. The series consisted of 2–10 individual whistles and had a maximum length of 2·5 s. Individual whistles and whistles forming part of series were of variable length ranging from 40 ms to about 1200 ms. Figure 49 shows that the length of individual signals did not fluctuate around a mean value as in a normal distribution, but that signals could be divided according to duration into discrete groups.

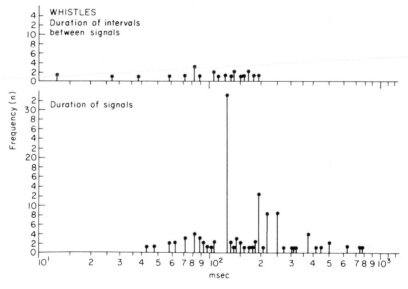

Fig. 49 Diagram of lengths of whistles recorded and intervals between whistles in the series analysed. For reasons of clarity a logarithmic scale has been chosen for the abscissa. The ordinate shows the number of measurements per class of length (class interval 5 ms).

A first group of very short signals had a duration of about 40–100 ms and showed considerable scattering. A second group was formed of signals of 125–150 ms. Most whistles in this group had a duration of between 125 and 130 ms. They were therefore almost of equal length.

The third group with a duration of 175–190 ms contained fewer signals. They again fell mainly into a single length class of 190 ms. At 215 ms duration, signals were to be found which all belonged to the same series.

There was a fourth group of signals of 250 ms duration which came from different series.

No other groups could be formed from the few whistles of longer duration. Figure 49 shows that the different groups occurred at regular intervals of about 60 ms.

The majority of all whistles recorded had a length of between 50 and 250 ms. Only a few signals lasted substantially longer. Individual whistles and whistles from series occurred in all length classes. Within a particular series probably produced by a single dolphin, the length of individual whistles could be either identical or might vary within certain limits.

The intervals between several whistles in a series were not correlated with the whistle length. Nor did the length of the intervals vary regularly within a particular series. Figure 49 shows that lengths of the intervals fluctuated within the same order of magnitude as the whistle lengths.

The majority of whistles recorded were frequency modulated. Short whistles were either simple of types ⌣, ⌣, or else ⌢ modulated. More complicated modulations could be broken up into a sequence of these three basic types as well as components of constant frequency.

C. Screams

When several dolphins were together in a group, whistles might turn into less frequently recorded sounds which we have named "screams". Screams were always grouped into series. These might consist either of screams alone or of a sequence of whistles and screams. Transitions from whistles to screams and vice versa occurred in the same series. The term scream covers a number of physically very different sounds. To the human ear on the other hand, they all sounded similar. What they had in common was several parallel frequency bands. In addition the individual sounds were not pulsed, in other words could not be further broken up into

single components separated in time. As in the case of the whistles, the
lengths of the individual sounds in a series fell into three groups as shown
in Fig. 50. The length of the intervals between different screams in a series
was of the same order of magnitude as for the whistles. It was also more or
less evenly distributed. Series of screams tended to last somewhat longer
than whistle series.

The only distinctive features which enabled one to separate screams and
whistles during analysis lay in the frequency domain. The upper frequency
limit of the screams in many cases reached about 30 kHz. The lower limit
at about 3 kHz lay at the same frequency as the whistles. Thus unlike the
whistles, the frequency range of the screams reached into the lower
ultrasound range. However, the highest frequency bands at 16-30 kHz
were never of maximum intensity. The most intensive frequency band
fluctuated within the same range as the whistles. Often two or three, and
more rarely four, distinctly harmonic frequency bands occurred. In the
case of long-drawn-out, strongly modulated screams, there was often a
single frequency band throughout. Other bands lying either below or
above were only occasionally visible on the sonagrams. Figures 51 and 42
show a typical situation in which maximum intensity occurs not at the
basic frequency but at a higher harmonic. The third harmonic is missing.
The fourth, weaker harmonic, does not show up on the sonagram but can
be recognized from the frequency analysis.

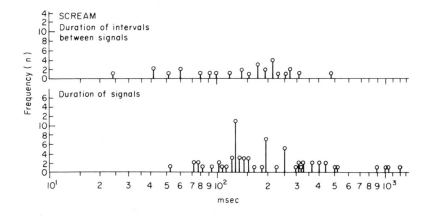

Fig. 50   Diagram of lengths of screams recorded and intervals between individual screams in the series analysed.

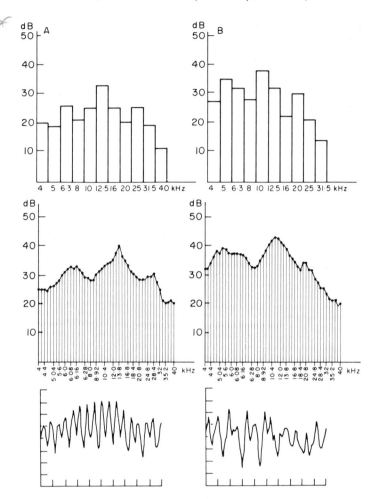

Fig. 51 Two screams from a series of 15 sounds. For purposes of comparison the same signal is shown with 1/3 octave spectra and narrow-band spectra (selective filter, $Q = 10$). Three distinct frequency bands can be seen, the uppermost one is no longer within the frequency range of the sonagram. Frequency spectra r.m.s., Oscillogram 125 µs/div.

Figure 42 shows the sonagram of the whole series of screams. It consisted of 15 individual sound emissions. The total length was about 3700 ms measured from the beginning of the first sound emission to the end of the last sound emission described.

1. Length about 180 ms. Frequency range 10–14 kHz, frequency- and amplitude-modulated sound. Synchronous with it a shorter sound with a 50 ms delay at a frequency range of around 6 kHz. Interval about 100 ms.

2. A short, 125 ms long, intensive sound at 10–14 kHz, frequency- and amplitude-modulated. Synchronous with it, with a slight delay, a softer sound component at 6–8 kHz. Interval about 110 ms.

3. A sound about 125 ms long, frequency range 10–14 kHz, frequency- and amplitude-modulated. Synchronous with it a signal component at 6–7 kHz, very similar to the second sound emission. Interval about 110 ms.

4. A sound about 140 ms long, frequency- and amplitude-modulated, frequency range 10–14 kHz. Synchronous with it a very short, barely indicated signal component in the 6 kHz range. Interval 140 ms.

5. A signal about 150 ms long, frequency range 9–15 kHz, frequency- and amplitude-modulated. Synchronous with it in the 6 kHz range a barely indicated, short frequency band.

These five screams are characterised by the fact that they are trough-shaped and consist of several frequency bands.

6.–8. There follows a longer pause of 550 ms. During this interval three groups of screams occur but they are very difficult to define.

9. A sinusoidal modulated signal about 440 ms long in the 10–15 kHz frequency range. At the beginning it appears to consist of two tones, 250 and 440 ms later respectively. (Superimposition of the sounds of several dolphins.) Interval about 40 ms.

10. This sound lasts about 340 ms and is sinusoidal and frequency-modulated. Amplitude modulation also occurs. Interval about 30 ms. Sounds of other dolphins again superimposed.

11. A short sound (about 140 ms) again trough-shaped, in a frequency range of 8–14 kHz, with clearly harmonic frequency bands. Between 4 and 6 kHz another short signal can be discerned. Interval about 45 ms.

12. Signal lasting about 130 ms in the 8 to about 16 kHz range with clearly harmonic frequency bands. Interval about 10 ms.

13. A very complicated signal lasting about 360 ms. At first sight it appears sinusoidally modulated. Closer examination shows that it consists of several frequency bands together in a range between

about 8–14 kHz. Synchronous with the first "trough" another sign can just be heard in the 6 kHz range, but it becomes clear only at the second trough. Interval 50 ms.

14. Trough-shaped whistle lasting about 120 ms at 8–15 kHz, with clearly harmonic frequency bands. Synchronous with it a very short signal component at 4–6 kHz. Interval about 80 ms.

15. A trough-shaped signal about 130 ms long. Frequency range 10–15 kHz. Synchronous with it, at about 6 kHz, a somewhat shorter frequency band.

Figure 52 shows the spectra of two screams from a series in the course of which the mean dominant frequency shifts upwards.

Figure 42 shows an example of a transition from screams to whistles in the course of a single series. The entire sound emission lasts 2630 ms. It consists of three groups of signals. The first signal is strongly modulated and has at least two harmonic frequency bands. The second signal is again modulated but considerably shorter. An additional frequency band is barely indicated. The series is followed by three typical whistles.

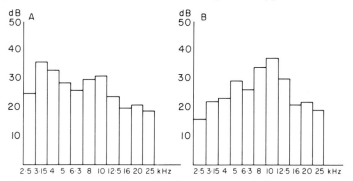

Fig. 52 Frequency spectra (r.m.s.) of two screams from a series of eight individual sounds. A: second sound in the series. B: fourth sound.

III. General Observations

Several facts emerge from the foregoing account of sounds emitted by *Sousa*:

1. The amount of respiratory air used during the emission of a train of whistles or screams lasting 10 s exceeds that of a train of clicks emitted by *Tursiops* by a factor 10^4.

2. If the capacity of the upper narial air chambers is comparable with that of *Tursiops* then they are quite inadequate to contain the total amount of air used and the excess must escape through the blowhole.

3. The non-pulsed, harmonic structure of the screams indicates that resonance occurs, that no recycling of air takes place during the scream and that thoracic pressure is involved.

4. It follows therefore that whistles and screams can only be produced whilst the animals are at the surface since the lungs must be presumed to be evacuated at depth not only from hydrostatic considerations but from anatomical observations of the structure of the lungs.

5. There can be no whistling or screaming at depth.

6. As the sonar pulses of *Sousa* were discernible at 200 m in the wild, those of *Delphinapterus* at 800 m, and calculated for *Tursiops* at 740 m, cetaceans must have considerable control over the power output of their sonar pulses since such amplitudes are never heard in Dolphinaria and perhaps for good reason.

Finally we may quote Bullock and Gurevich (1979) who stated after an extensive review of the literature on the subject:

"It is compatible with a view that dolphins are not vastly different from other nonhuman mammals in having a limited variety of signals with distinct information content, probably much less than 40, among which some are complex but may not carry much information beyond identifying the individual and a crude indication of its state. There is no compelling suggestion of syntax or vocabulary at a level beyond that common in non-primate orders".

V.   Sonar Fields (*Inia geoffrensis*)

Most of the acoustic observations on delphinids described in this book were carried out on the more exotic, endangered species during the course of ecological studies and consequently those more readily available in dolphinaria, have been omitted. However, the Amazonian river dolphin *Inia geoffrensis*, although belonging to a different family, can be considered acoustically to be fairly typical of the majority of the delphinidae.

Observations were made on three specimens of the Orinoco race of Inia (*Inia geoffrensis humboldtiana*) from the Rio Apure which had been brought back to the Duisburg Zoo (FRG) in 1975. At the time of the measurements they had already been in captivity for 2 years (see Pilleri *et al.*, 1979).

I. Methods

The three dolphins, one adult and two sub-adults, were kept in a concrete pool measuring 7 × 5·8 m, the water being usually 1·7 m deep. An underwater viewing window made observation possible from the side. The 8103 or 8101 Bruel and Kjaer recording hydrophone was installed in a corner of the pool at a depth of 90 cm or 1·2 m respectively, and the transducer was positioned 0·8 or 1 m away from the two side walls. This arrangement ensured that every time the dolphins swam round, they advanced directly towards the hydrophone and turned away from it at a distance of about 1 m. All the measurements were made on free swimming and non-conditioned animals. One observer sat behind the hydrophone on a platform above the water. From there he could watch the dolphins from above and record his remarks on tape. The second observer filmed the situation from above or through the underwater viewing window.

Figure 53 shows the equipment used for the actual recording of the dolphin sounds and the subsequent analysis in Berne. The entire system had a bandwidth of 0·1–150 kHz (± 3 dB) at a recording speed of 30 ips. The recording unit consisted of Bruel and Kjaer hydrophones, pre-amplifier, measuring amplifier, and a Stellamaster tape recorder (Stellavox, Neuchâtel) specially adapted to four-track instrumentation use. Channels 1 and 2 were used to record the dolphin signals. Channel 2 was set to have a maximum recording level 20 dB higher than channel 1. This gave a total useful dynamic range of 55 dB. Channel 3 was used for the commentary and channel 4 for recording the synchronization signals of a Super-8 cine-camera.

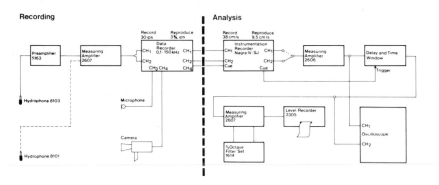

Fig. 53   Block diagram of the electronics used for recording and analysis.

In the laboratory the signals were transposed at one-eighth speed to a Nagra IV-SJ (Kudelski) measuring tape recorder. Loops were then cut from these tapes to facilitate subsequent analysis and a rectangular time window with adjustable time delay was used to extract single sonar clicks for analysis. A window of 160 μs (real time) was used for all the analyses. Oscillograms were plotted from the clicks selected for analysis; and spectro-grams were made with the help of the 2607 (peak detector with 30 μs rise time) and the 1614 1/3 octave filter set. In order to ensure an optimum signal-to-noise ratio we analysed weak signals from channel 2 and louder signals from channel 1 in case channel 2 was saturated.

The following criteria were used for selection of the sound recordings and the corresponding film sequences to be analysed. Requirements to be met by the recorded sound sequence were:

1. Only the one dolphin swimming towards the hydrophone was emitting sonar sounds.
2. The dolphin continued to emit sounds during the whole sequence of movement to be investigated.

Requirements to be met by the film scene:

1. The position of the dolphin in relation to the hydrophone was clearly defined and constant in either the vertical plane, or the horizontal plane—in other words an oblique position or oblique turning away during the sequence was not acceptable.

The horizontal (lateral) extension of the emission field was determined from the three dolphins swimming freely round the pool when they usually swam one behind the other in regular counter-clockwise circles. Data for the polar diagrams were always evaluated from the same sequence: the dolphin swimming up to and turning past the hydrophone at a distance of about 1 m with the hydrophone positioned at the swimming level of the dolphin (about 80 cm above the bottom). Strict application of the above-mentioned criteria meant that out of a much larger number of possible situations only five sequences remained valid for determination of the horizontal polar pattern.

The vertical (dorso-ventral) extension of the emission field was determined by means of another behavioural situation in which we made use of the dolphins' curiosity about unfamiliar objects in the water. Behind the recording hydrophone which the dolphins had become used to, we moved an object up and down—which never failed to attract the animals' attention. Sometimes one or more dolphins would remain motionless in front of the hydrophone, moving their heads up and down to follow the moving object (Fig. 54). This behaviour was filmed through the underwater window while the observer above the pool attempted to move the object within the vertical plane formed by the hydrophone and the rostrum axis of the dolphin. Any deviations from this plane were noted on the tape. Once again application of the selection criteria meant that out of a large number of similar situations only four sequences were in fact suitable for analysis.

The selected film sequences were analysed frame by frame. The angle of the rostrum axis to an imaginary line joining the corner of the mouth to the distance of the animal from the hydrophone and for the dorsal-ventral sequences the position of the object being inspected. With the help of the tape synchronization signals it was then possible to evaluate for each individual recorded sonar click the corresponding position in the sound field. Thus the measured angles could be assigned to the 1/3-octave spectra of single clicks shown in Figs 55 and 56. Assuming that the signal emitted by the dolphin during the recorded sequence did not change, the measured changes in amplitude of an individual 1/3 octave band represented the variation of the sound pressure at that frequency as a function of the changing relative position of the hydrophone in the sound field of the moving dolphin. The polar diagrams shown in Figs 57 and 58 were obtained by plotting the relative 1/3 octave values in dB of a specific frequency against the angle of the dolphins' rostrum axis to the hydrophone.

II. Object-locating Behaviour

New objects introduced into the pool were circled by the dolphins and investigated from different sides. A dolphin would remain motionless for some time in front of an unknown object and if we moved the object the dolphin would follow it with its head and in this event the rostrum axis was always pointed directly towards the object being investigated (Figs 54 and 62). During this behaviour sonar signals were often, but not always, audible. Familiar objects seldom provoked the same reaction, they were briefly located and afterwards ignored.

There is no doubt that the *Inia* in the clear water of the pool used their eyes as well as the sonar signals. This was particularly obvious when the dolphins were given objects to play with. While waiting for a brush or a ring to be thrown for them, the dolphins often swam close to the observer and followed his movements which took place outside the water.

III. Analysis of Sonar Clicks

The recorded sonar signals consisted of a series of short, broad-band pulses (clicks). The pulse trains of a motionless animal were very uniform, in other words individual clicks could scarcely be distinguished from one another. Slight changes in the position of the animal, however, resulted in a considerable alteration in the characteristics of successive signals.

Figures 55 and 56 show two examples of signal sequences during a monitored change in the dolphin's position. Figure 55 indicates the change

Fig. 54 Echolocation of objects by the Orinoco dolphin (*Inia geoffrensis humboldtiana*). The animals follow a brush moved up and down behind the hydrophone (arrow) with movements of the head.

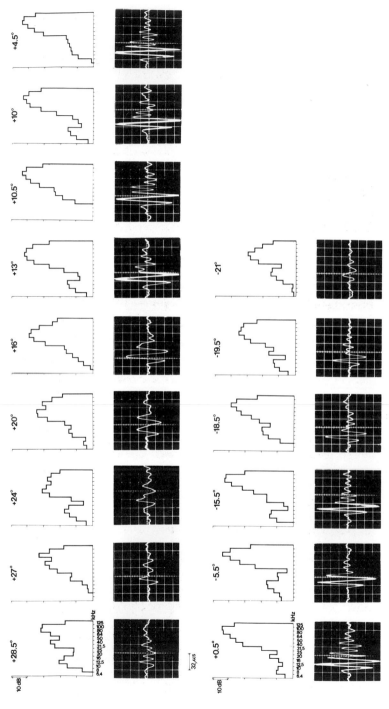

Figs 55–56 Frequency spectra and oscillograms of individual clicks from two pulse trains in the course of which the dolphin changed its position in relation to the stationary hydrophone. **Fig. 55**: Signal pattern during movement of the dolphin's head in the dorso-ventral plane. The dolphin was following an object moved up and down behind the stationary hydrophone with vertical movements of the head. In the case of each click plotted, an indication is given of the angle between rostrum axis and hydrophone at the time of recording. Rostrum axis = 0°.8103 recording hydrophone.

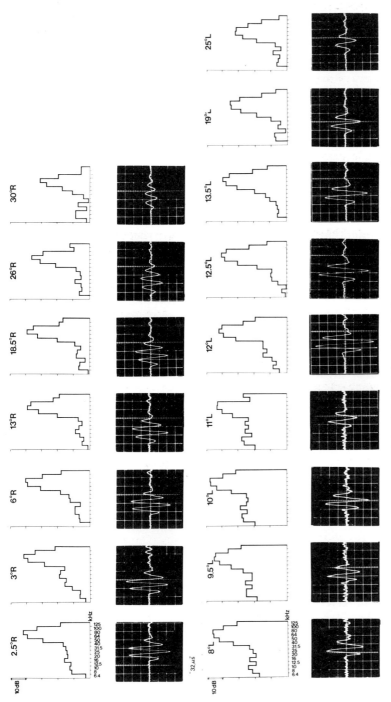

Fig. 56 Signal pattern during a movement of the dolphin's head in the horizontal (lateral) plane. Sixteen out of a total of 34 clicks are shown. R = hydrophone seen to the right of the rostrum axis, L = hydrophone to the left of the rostrum axis. The clicks shown from 8–11° on the left originate from channel 2 of the tape recorder with a higher recording level, as channel 1 was over-modulated by these signals. The signal spectrum was masked by noise in the frequency range below 25 kHz. In the corresponding oscillograms the amplifier gain was 6 dB lower than in plotting the other clicks. 8101 recording hydrophone.

in the signal with the hydrophone displaced in the median (dorso-ventral) plane, while Fig. 56 shows the signal change with the hydrophone displaced in the horizontal (lateral lateral) plane through the rostrum axis. The frequency spectra show that only at frequencies above about 25 kHz was the siganl clearly discernible above the water noise. The absolute broadband sound pressure level (0·1–200 kHz) was 37 ± 5 dB peak re. 1 μbar at 1 m. Individual clicks in the axial part of the emission field (around 0°) had a dominant frequency of about 80 kHz. The 1/3 octave spectrum had a single peak and reached as high as 150 kHz, to the upper frequency limit of our recording equipment. The pulses had 2–3 periods and were 40–60 μs long. Towards large angles, the level of the signals decreased rapidly and at the same time the dominant frequency of the clicks shifted downwards. This meant that high frequencies were more narrowly beamed than lower frequencies. Due to the lowering of the pulse frequency, the duration of peripheral signals was longer and the rise time was slower than in the clicks recorded near the axis of the rostrum. Figures 55 and 56 show only every second to fourth click in a train. The transition between characteristics in successive clicks was continuous. Thus single clicks are considered to be representative of the sound properties at the corresponding position in the sonar field.

In the median plane of the emission field, in the case of intensive signals recorded in the central region of the field, high-frequency hangovers (100–125 kHz) were to be observed. These hangovers contributed essentially to the domination of high frequencies shown up in the frequency spectra. They were more or less prominent in all the signals recorded in the dorso-ventral plane of the field (8101 recording hydrophone). Clicks recorded in the horizontal plane through the emission field did not show this kind of resonance (8101 and 8103 recording hydrophones). This was true irrespective of which hydrophone was used. Nevertheless, more recordings would be needed in order to rule out with certainty any possible technical cause of the phenomenon.

IV. Directivity of the Emission Field

Polar diagrams (Figs 57, 58) of the emission field in the median and horizontal plane were plotted on the basis of Figs 55, 56 and with the help of similar data not reported here in detail. It was assumed that *during the sequences* analysed, lasting 1–2 s in each case, the dolphins produced clicks

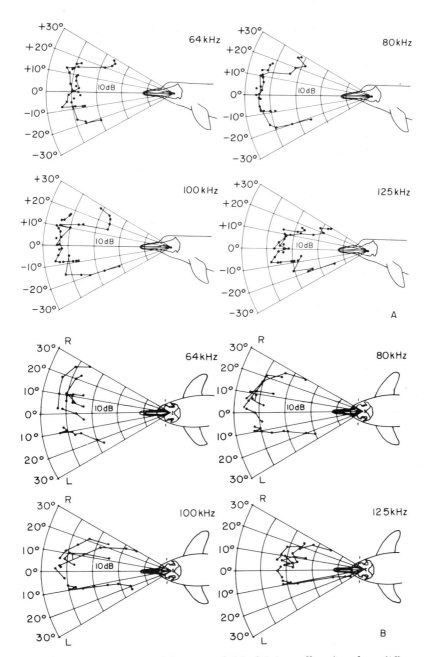

Figs 57–58 Polar diagrams of the sonar field of *Inia geoffrensis* at four different signal frequencies, based on 1/3 octave peak value measurements. A: dorso-ventral section; B: lateral section. Measurements of sonar clicks in the same pulse train are connected by lines. The origin of the coordinate system lies in the region of the larynx. The prolongation of the rostrum axis forms the 0° line.

with unchanged amplitude and frequency. This assumption appeared to be justified as the clicks from a particular region of the sonar beam in a number of recording situations involving different behavioural contexts were very similar. Moreover, when the dolphin remained in the same position, brief level fluctuations within a pulse train occurred only infrequently.

The median (dorso-ventral) and the horizontal extension of the emission field were approximately the same. The central axis of the beam corresponded with the axis of the rostrum ($0°$).

One major characteristic of the beam at any of the traced frequencies was a relatively wide maximum intensity in the axial region and a steep falling off in intensity on all sides towards the periphery. High frequencies appeared to be more narrowly beamed than low frequencies (Table 5).

A. Median plane through the sonar field

Table 5:   Median plane through the sonar field.

Frequency (kHz)	Dorsal	-3 dB	Limit Ventral	Beam width
64	$16°$		$19°$	$35°$
80	$13°$		$16°$	$29°$
100	$12°$		$16°$	$28°$
125	$11°$		$12°$	$23°$

Several differences were to be observed between the dorsal and the ventral sectors.

B. Ventral sector

Maximum intensity reached somewhat *further ventrally of the rostrum axis than dorsally*. The 80 kHz frequency had the highest intensity, 100 kHz was only slightly lower (2 dB) whereas 64 kHz was 4 dB lower and 125 kHz about 15 dB lower. At 64, 80 and 100 kHz the sharp falling off in intensity began at $-15°$ to $-20°$, while at 125 kHz it began already at $-10°$.

C. Dorsal sector

At all four frequencies an intensity drop of about 15 dB was to be observed between +10° and +20°. At 80 and 100 kHz signal intensity began to rise again at angles greater than 30°. This led, in conjunction with the different width of the dorsal and ventral sectors, to a slight asymmetry of the emission field with respect to the rostrum axis.

D. Horizontal plane through the sonar field

In the horizontal plane the field at all frequencies was only approximately centred round 0°. Asymmetry was more marked than in the median plane. The intensity maximum was about 10° broader to the right than to the left. The frequency of 80 kHz showed the highest intensity, while at 64 kHz intensity was slightly lower and at 100 kHz and 125 kHz it had fallen off by as much as 4 and 15 dB respectively. Sixty-four, 80 and 100 kHz showed a steep drop in intensity to the right of 20° whereas to the left intensity already began to fall off at about 10° (Table 6).

Table 6: Horizontal plane through the sonar field.

Frequency (kHz)	−3 dB limit		Beam width
	Left	Right	
64	12°	29°	41°
80	11°	17°	28°
100	10°	16°	26°
125	11°	18°	29°

It was stated by Pilleri *et al.* (1979) that the shape of the sonar field in *Inia* is primarily related to the anatomical structure of the dolphin's head. The peripheral limits of the sound beam are caused by a number of air-sac systems (Fig. 59) from whose surface the sound waves are reflected. Ventrally, the palatine lobes of the pterygoid sinus in *Inia* prevent exit of the signals. Laterally, the large orbital lobes of the pterygoid sinus perform the same function. As injection preparations showed, the orbital lobes of the pterygoid sinus reach right up to the dorsal cranial plane. Dorsally, three superimposed paired air-sacs, the nasofrontal, the vestibular and especially the premaxillary sacs prevent the sound from leaving the head.

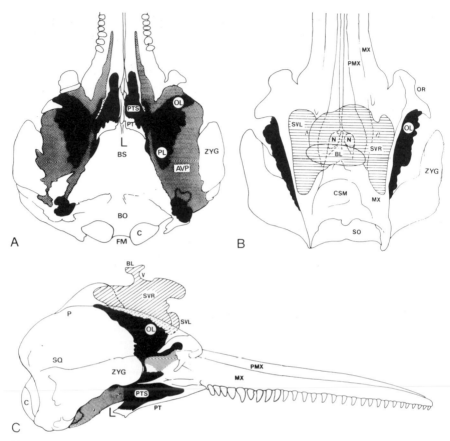

Fig. 59 A: ventral, B: dorsal and C: lateral view of the nasal tract complex and accessory air sinus of the Amazon dolphin *Inia geoffrensis*. AVP = arterio-venous plexus (rete mirabile), BL = blowhole, BO = basioccipitale, BS = basisphenoid, C = condyle, CSM = supramaxillary crest, FM = foramen magnum, L = larynx, MX = maxilla, N = naris, OL = orbital lobe (of the pterygoid sinus), OR = orbit, P = parietale, PL = posterior lobe (of the ptery-goid sinus), PMX = premaxilla, PT = pterygoid bone, PTS = pterygoid sinus, SO = supraoccipitale, SQ = squamosum, SVL = left vestibular sac, SVR = right vestibular sac, V = vestibule, ZYG = zygomaticum.

The vestibular sac in *Inia* is very large and reaches further forward than the other two sacs. When inflated, these nasal diverticula reach the outer edge of the maxilla on either side and extend forwards as far as the orbita.

Altogether the system of air-sacs forms a screening area around the larynx with the result that sound can only be emitted in a forward direction.

The sonar field in *Inia* extends in the median plane from $+30°$ to $-20°$ around the axis of the rostrum. These limits roughly correspond with the angle formed between the dorsal edge of the palatine lobes of the pterygoid sinus ($-20°$) and the rostral end of the vestibular sac ($+50°$). The dorsal limit of the palatine lobe of the pterygoid sinus—recognizable as the boundary line between the palatum and the maxillare—runs at an angle of $-20°$ to the row of teeth in the rostrum (see Pilleri and Gihr, 1979: Fig. 17). The dorsal limits of the sonar field cannot be as accurately determined on anatomical grounds as the ventral and lateral boundaries, because the vestibular and premaxillary sacs are not surrounded by bony structures

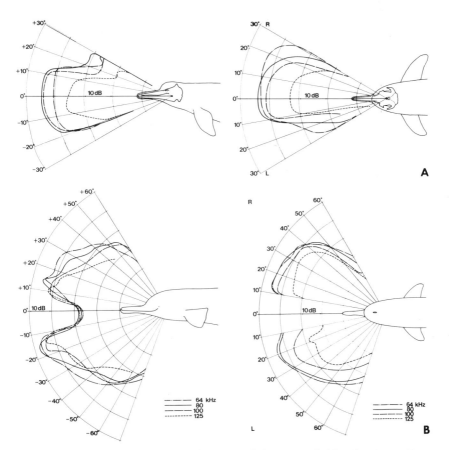

Fig. 60 Comparative schematic diagram of the sonar fields of *Inia geoffrensis* and *Platanista indi* (after Pilleri *et al.*, 1976). A: Dorso-ventral section, B: lateral section.

and are flexible as a result. Different degrees of inflation of the sacs could give rise to changes in the dorsal limit of the emission field. We can neither confirm nor deny this possibility on the basis of our recordings as we had too little suitable material available for assessing the constancy of the limits of the sonar field. The shape of the field described here should thus be regarded as a preliminary evaluation of the directional characteristics of the sonar signals.

As already discussed by Pilleri *et al.* (1976), *Platanista* presents an entirely different picture anatomically, and consequently has a very different shaped sonar field. The most striking anatomical characteristic of the head of *Platanista* is the two pneumatized cristae maxillares which rise like a helmet above the viscerocranium (Purves and Pilleri, 1973–1974, 1978). The powerful, paired maxillary sinus which is connected with the middle ear cavity is responsible, in conjunction with the premaxillary sacs, for the dorsal and partly also for the lateral limitation of the emission field. The vestibular sac is missing in *Platanista*. The pterygoid sinus stretches far back, stopping just short of the bulla tympanica, and extends laterally upwards to the edge of the maxillary crests, forming the ventral and ventro-lateral screening for the sound source. Unlike the situation in *Inia*, the air-sacs responsible for screening are closely connected with the bony structures and allow no marked changes in the limits or shape of the beam.

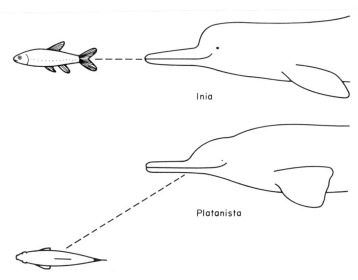

Inia

Platanista

Fig. 61 Angle of the rostrum in *Inia geoffrensis* and *Platanista indi* when locating a fish. *Inia* swims on its belly, *Platanista* on its side.

According to Pilleri *et al.* (1976) the experimentally determined sonar field of *Platanista* (Fig. 60) consists of two sound beams of different origin. The dorsal beam is limited on the one hand by the edge of the maxillary sinus and on the other hand by the pterygoid sinus and nasolaryngeal air-sacs. The beam is formed by direct signals emerging from the larynx. This part of the sound field corresponds with the entire emission field in *Inia*. The ventral limit in *Platanista*, however, is displaced so far upwards

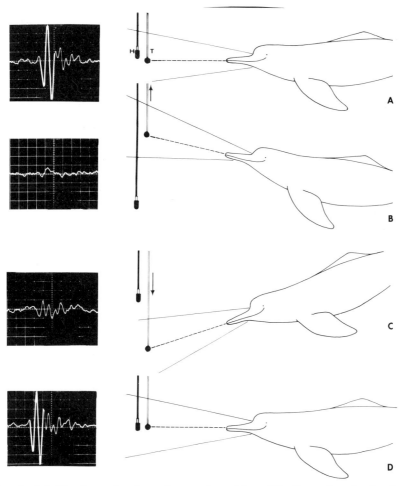

Fig. 62 (*A–D*) Investigation of a moving object (T). Recording hydrophone (H). Maximum sound intensity was recorded in the region of the rostrum (A, D). Ventrally (B) and dorsally (C) of the axial region round the rostrum, signal intensity fell off rapidly.

that the central axis of the direct sound field lies about 15° above the line of the rostrum. At the level of the rostrum axis the sound pressure is already 20–25 dB lower than in the field axis. In *Inia*, on the other hand, field axis and rostrum axis are one and the same. In *Platanista* an additional ventral sound beam is formed from signals reflected by the cristae maxillares. No homologous field exists in *Inia*. *Platanista* uses only the ventral sound field for locating and inspecting prey or unfamiliar objects (Pilleri *et al.*, 1976), that is to say the objects were located at an angle of −30° to the rostrum axis (Fig. 61). *Inia*, on the other hand, directed the rostrum axis towards any object being located and investigated (Figs 62 and 54). The different behaviour of the two species is therefore due to the differing beam patterns.

VI. Properties of the Sonar System of Cetaceans with Pterygoschisis: *Delphinapterus leucas, Neophocaena phocaenoides, Phocoena phocoena*

In the majority of the Delphinidae the pterygoid hamuli of the skull are greatly pneumatized and elongated antero-posteriorly as shown in the Fig. 63A. These air spaces in life are normally filled with an oil mucous foam and it was explained on p. 121 why these would be effective in channelling sounds produced by the epiglottic spout to the bones of the rostrum by way of the palatopharyngeal sphincter. It was also explained why sounds produced in the lower part of the larynx (probably of lower frequency) would not be conveyed to the rostrum.

However, in a series of papers Pilleri (1979–1981) drew attention to the fact that in many species such as the White whale, *Delphinapterus* (Fig. 63B), *Monodon*, the narwhal, and members of the porpoise family *Phocoena* (Fig. 63C) and *Neophocaena* (Fig. 63D), as well as some of the Delphinidae e.g. *Cephalorhynchus*, the pterygoid hamulus was only partially pneumatized and that there was a wide gap between the hamuli of each side. To this he gave the name "pterygoschisis" and assumed that it would have an effect on the ventral distribution of sound and possibly on its frequency structure. In the event, these predictions turned out to be true.

Whilst carrying out the observations on *Sousa* previously described he was also able to make records of the sounds emitted by the Finless porpoise *Neophocaena phocaenoides*. The records were made between December and January 1973, 1974 in the mangrove swamps of the Indus Delta. Three

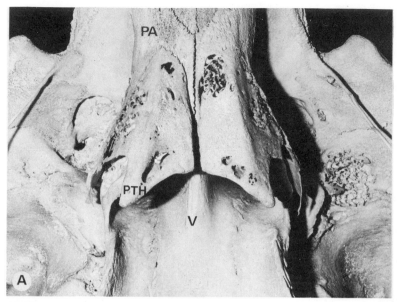

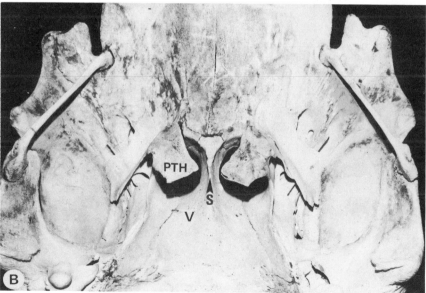

Fig. 63 Basal view of the skull of A: *Tursiops truncatus*, B: *Delphinapterus leucas*, C: *Phocoena phocoena*, D: *Neophocaena asiaeorientalis*. PA = palatinum, PTH = pterygoid, S = pterygoschisis, V = vomer.

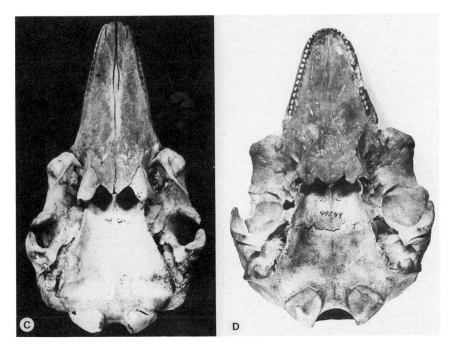

specimens had been caught with a net and kept in a closed-off section in the Kai Creek. In order to obtain information about the directional radiation of the sonar sounds, one of the porpoises was held underwater and the sounds were recorded with the hydrophone placed at different positions around the head, very close to the surface of the skin. After each measurement the porpoise was raised to the surface to breathe. When the experiments were completed, all three specimens were released. These records were not analysed until after 1978 when the pterygoschisis hypothesis had been developed.

In the spring of 1978 when the sound records of *Inia geoffrensis* were made (Pilleri *et al.*, 1979) the opportunity arose to make sonar observations on the White whale, *Delphinapterus leucas* (Zbinden *et al.*, 1980).

I. *Delphinapterus leucas* : Methodology

The recording system consisted of an 8101 Bruel and Kjaer sensitive hydrophone, a 2607 B and K measuring amplifier and a Kudelski S.A. Nagra IV-SJ measuring tape recorder.

The hydrophone had a nominal sesitivity of 580 μV/Pa and a flat frequency response up to 85 kHz (−3 dB). The measuring tape recorder limited the bandwidth of the recording system to 35 or 40 kHz (−3, or −5 dB). For the laboratory analysis of the sounds the equipment already described by Pilleri *et al.* (1979) was used. For single click analysis in 1979 the authors had been using a rectangular time window. They replaced that window by a 5623 B and K Gauss impulse multiplier. This was necessary because the White whale at the high repetition rates commonly observed in the low frequency components of single sonar clicks merged to form a continuous signal. Using rectangular time weighting on a continuous signal would have given rise to severe distortion of the signal frequency spectrum. In order to minimize distortion it was convenient to fade the signal in and out with a gaussian rather than a rectangular weighting function. Using a Gauss impulse multiplier half amplitude width of 25 ms (real time) and with 1/3 octave filters, it was possible to determine the spectra accurately down to a lower limit of 125 Hz. Below this frequency, with the above cited window length, the effective measuring bandwidth no longer depended on the bandwidth of the filter but could be determined uniquely by the window function.

For a first rough estimate of the position and structure of the emission field the whale's acoustic location of prey was observed. This meant that they had to place squid, their normal diet, in a direct line between the whale and the recording hydrophone.

A trainer with whom the whales were familiar enticed them away from the hydrophone with the help of food and another squid was then thrown into the water behind the animals' backs and directly in front of the hydrophone. The whales immediately turned about and swam towards the squid, which meant that they were also swimming towards the hydrophone. The entire behaviour sequence and the position of the whales in relation to the hydrophone was noted during the recording on a second track on the tape and different stages in the behavioural sequence were indicated with time marks.

The process was repeated a number of times with the hydrophone hanging at different depths. As the whales swam towards the hydrophone in varying positions, with the squid sometimes in front of, and sometimes under, above, or to the side of the hydrophone, recordings were obtained from all possible areas of the front part of the head. This kind of method did not, of course, produce precise, quantitative data on the structure of the emission field, it could only give a qualitative description of the main directions in which the sonar signals were propagated. At the same time we were able to establish certain relations between the emission field and the anatomical structure of the head.

II. Emission Field

The fact that the whales' eyes are directed obliquely forwards and downwards means that they have the best view of the surroundings from this position (Fig. 64). The whales were able to follow the movements of people near the pool and react accordingly. Even underwater they probably use their eyes for locating large objects. On a number of occasions a whale was observed swimming up to the hydrophone from below and staring at it for a moment with one eye.

One of the three whales took a great interest in the hydrophone soon after the instrument was first placed in the water. It often swam straight towards the hydrophone, (Fig. 65) touched it gently with the melon and pushed against the wall of the pool, with the melon becoming elastically

Fig. 64   Rostro-ventral position of the eyes in the white whale.

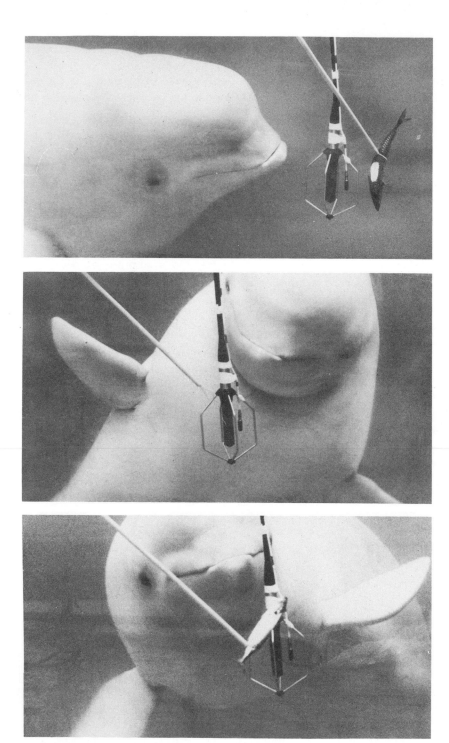

Fig. 65 White whale approaching and acoustically investigating the hydro-phone attracted by food.

deformed in the process. This behaviour was repeated several times after which the animal allowed the hydrophone to glide along its side. Later we once or twice saw the whale taking the hydrophone in its open mouth while lying on its side.

III. Sounds Uttered while Swimming Towards Prey and Unfamiliar Objects

While the whales were swimming freely about in the pool, only occasionally weak pulse sounds could be heard when the animals were close to the hydrophone. Even when a whale was swimming straight towards the hydrophone, click sounds only became distinctly audible when it was less than 2 m from the hydrophone. At the same time the impression was that the clicks were most frequently audible when the whale turned the front part of the throat in the direction of the hydrophone.

While swimming towards the hydrophone, or a fish (Fig. 65), the whales produced rapidly repeated clicks which, when listened to during the recording, sounded rather like the noise made by a rusty hinge. However, when played back on the tape at reduced speed, the individual clicks were clearly distinguishable. The laboratory analysis revealed two frequency ranges (Fig. 66).

One range with a dominant low frequency of 0.5-2 kHz was entirely audible to the human ear. The second range with a dominant frequency of 30-40 kHz lay at the upper limit of our recording equipment and was therefore not audible to the observer during recording.

The low-frequency and high-frequency components (LF and HF components) were synchronized with each other in time. In the oscillograms the HF component showed up as a sharp click, superimposed at a certain point on the considerably longer LF pulse (Fig. 67).

The LF pulse was 2-4 ms long and showed 2-4 periods. Although the pulse shape was very variable, the dominant frequency and bandwidth always remained approximately the same. The bandwidth ranged from 100 Hz to 4-8 kHz. In this range the frequency spectrum contained both components of the individual LF pulses and components of the repetition rate of the click train. In the recordings the superimposed HF clicks varied in intensity depending on the position of the emitting whale. In cases with a strong HF component the spectral component of the repetition rate in the LF range was often dominant over the frequency

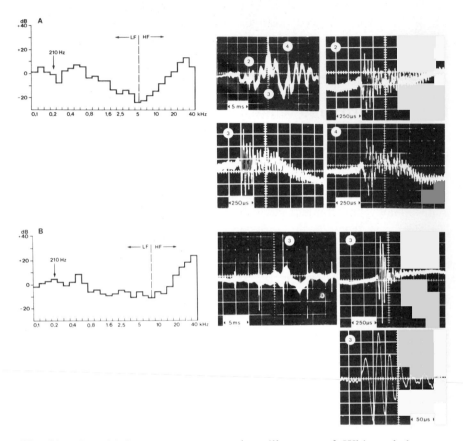

Fig. 66. One-third octave spectra and oscillograms of White whale sonar signals. In the frequency spectra low-frequency (LF) and high-frequency (HF) components are to be distinguished. The repetition rate of the HF clicks was 210 Hz in both A and B. In the general oscillograms belonging to analyses A and B the influence of the gaussian window (12·5 ms half amplitude width) is clearly visible as a rise in sound level on the left and a symmetrical fall in level on the right. A: Series of 5 clicks with fused LF components. The consecutive HF clicks 2, 3 and 4 show marked hangovers. Dominant frequencies 0·5 kHz (LF) and 31 kHz (HF). B: Another series of 5 clicks with fused LF components. The HF component is considerably stronger and higher in frequency compared with A. Dominant frequencies 0·5 kHz (LF) and 40 kHz (HF click No. 3 is shown in various time division settings).

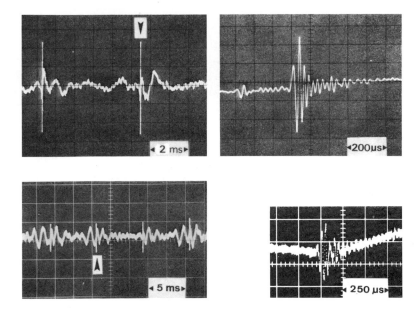

Fig. 67   At a lower repetition rate (in these two examples 190 Hz and 160 Hz) the LF clicks are clearly separated from each other and can be followed synchronously with the HF click train. In both series a single HF component (arrow) is shown on an expanded time base.

component of the LF pulse. The position of such peaks and their harmonics varied according to the repetition rate and to a considerable extent determined details of the spectrum pattern in the LF range.

The length of the superimposed HF clicks varied between 75 and 250 μs ($\bar{x} = 170$ μs). Changes in length were due on the one hand to different number of periods or different frequency (Fig. 68) and on the other hand to the presence of hangovers* (Fig. 66). The shortest clicks observed had a length of 2.5 periods. Normally the number of periods was 3 to 5. All the HF clicks recorded were frequency modulated. The first distinct half-wave always had a lower frequency than the following waves. The frequency increased towards the middle of the pulse and, in the case of longer pulses, declined again in the last half-wave (cf. Figs 66, 68 and 71).

---

* Reverberation and reflection of sound at the tank walls or the water surface can be ruled out in considering the origin of the signal-tails which we called hangovers. The phenomenon may originate in pulse stretching due to radiation from an extended surface at wide angles to the propagation axis and in multipath travelling of sound waves within the head of the dolphin.

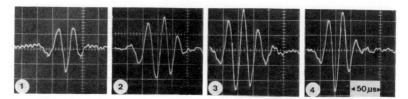

Fig. 68 Examples of HF clicks from the axial region of the emission field. Note the frequency modulation of the signal.

Table 7: Frequency modulation of the HF signal component

Figure/ Click No.	kHz (+ 1 kHz)										No. of periods
	1.	2.	3.	4.	5.	6.	7.	8.	9.	10. Half-wave	
Fig. 3/1	33	33	42	42	48	—	—	—	—	—	2·5
Fig. 3/2	33	33	37	38	45	42	—	—	—	—	3
Fig. 3/3	33	37	37	42	42	42	33	—	—	—	3·5
Fig. 3/4	33	37	42	39	48	42	—	—	—	—	3
Fig. 1B	31	33	40	44	50	50	44	—	—	—	3·5
Fig. 6/A14/5	33	38	38	38	42	38	42	36	38	33	5
Fig. 6/A15/4	31	42	40	42	40	40	42	38	36	35	5
Fig. 6/A15/12	35	36	40	40	42	40	40	38	38	32	5

In the frequency spectrum the HF click component was normally visible above the noise level at about 8–10 kHz and reached a maximum at 30–40 kHz.

In individual cases with very low background noise a rise in the spectrum could be found already at 4 kHz, at the upper limit of the LF component.

In general the repetition rate of the click trains was particularly high just before contact with the object being investigated. At high repetition rates the low-frequency pulses merged to form a continuous vibration on which the series of high-frequency clicks was superimposed. Repetition rates of up to 400 Hz were measured when the tip of the rostrum of the approaching whale was 10 cm from the hydrophone. On average the repetition rate when the whale was closer than about 1 m was 250–350 Hz.

IV. Emission field of *Delphinapterus*

Figure 69 shows a diagram of the profile of the head of a white whale. Signals recorded from different positions appear in the bar diagrams around it. They indicate only the basic structure of the sonar emission field. A selection of typical signals is shown in Fig. 70 A–C.

To make interpretation of the data easier, it shall first compare the signal intensity at different recording positions. This provides information on both the general direction of radiation of the high- and low-frequency signals, and the axis of the acoustic field. In the second place we shall

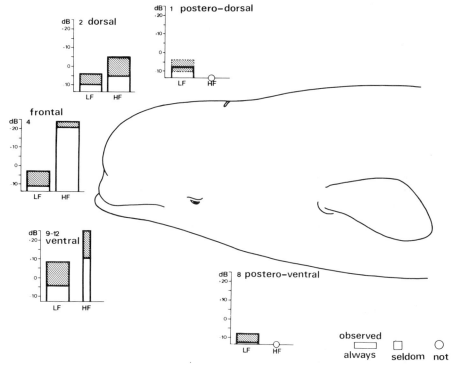

Fig. 69 Diagram of signal characteristics at different areas of the sonar field. The sound pressure level of the LF and HF components is shown in bar form in dB re 1 μbar (1/3 octave at 1·6 and 40 kHz). Corresponding broadband sound pressure levels (0·1–35 kHz) can be found in the text. The hatching indicates the area between maximum and minimum sound pressure level found in each sector of the field. The bar widths show how often the LF and HF components were observed in the checked recording sample. Numbers 1–12 refer to the corresponding signals in Fig. 70 A.

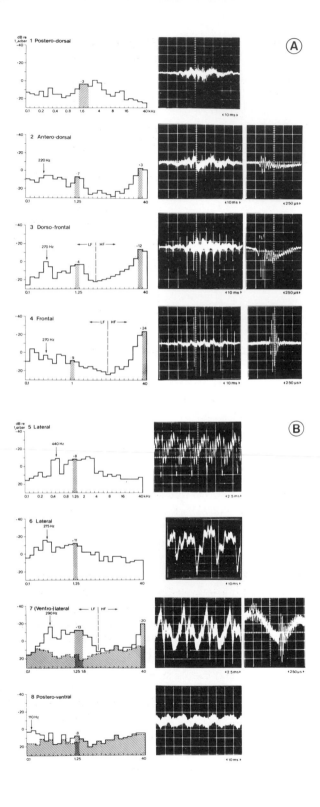

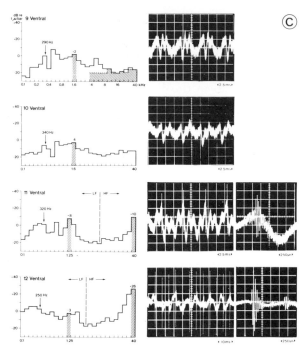

Fig. 70 (A–C) One third octave frequency spectra and corresponding oscillo-grams of sonar signals. The numbers of the analyses refer to the position of the hydrophone in the sonar field as shown in Fig. 69. The repetition rate of the click trains is indicated in the spectrum by its value in Hz and an arrow. Dominant frequencies are indicated with hatching and level data (dB re 1 μbar). The half-amplitude width of the gaussian window used for analysis of the photographed signal section was 25 ms. In all signals with a HF component the general oscillo-gram is accompanied by a single pulse shown on an expanded time base. A: Typical signals from the dorsal to frontal sector of the sonar field. Analysis 1: No HF component. The frequency ranges 3–5 kHz and 10–12 kHz are probably masked by water noise. Analysis 2: Fairly strong HF component. The HF click shows hangovers. Analysis 3: Strong HF component. The HF click shows hangovers. Analysis 4: Strongly dominating HF component. The HF click is simple, without marked hangovers. B: Typical signals from the lateral and posteroventral sector of the sonar field. Analysis 5/6: No HF component. Marked LF component. Above 5 kHz the signal is masked by background noise. Analysis 7: Strong HF component as well as marked LF component. In the spectrum the background noise is shown oblique-hatched. Analysis 8: No HF component. Signal hardly visible above the oblique-hatched background noise. C: Typical signals from the ventral sector of the sonar field. Analysis 9/10: No HF component. Fairly strong LF component. Above 8 kHz the signal is masked by background noise (oblique-hatched). Two examples of a ventral signal outside the narrow HF ventral field. Analysis 11/12: Strong and domina-ting HF component with fairly strong and strong LF component. The HF clicks have only small hangovers.

indicate how often the LF and HF signal components occurred in the recording samples as a function of the recording position. Thirdly, we shall draw conclusions about the directivity of the LF and HF fields, and lastly we shall discuss some special characteristics of signals recorded from known positions.

A. Comparison of Intensities

Broad-band levels in dB re 1 µbar were measured from oscillograms. Recording distance was less than 1 m. The observed fluctuations of up to 10 dB must be considered in the light of the simple experimental setup. Changes in distance between dolphin and hydrophone and changes in the exact angular position of the animal with respect to the transducer between trials may explain the high fluctuations we measured.

(a) Intensity of the low-frequency component was greatest in lateral signals (15–20 dB), somewhat less in ventral signals (5–15 dB, and considerably less in dorsal and frontal signals (-5 to $+$ dB).
    LF   Lateral > ventral > > dorsal = frontal

(b) Intensity of the high-frequency component was greatest in frontal signals (20–25 dB), somewhat less in ventral (15–25 dB) and lateral (20 dB) signals, substantially less in dorsal signals (0–5 dB) and lacking in ventral signals recorded at the base of the flipper.
    HF   Frontal $\geq$ ventral   ($=$ lateral) > > dorsal

(c) Comparison of HF and LF intensities shows that in all cases with an HF component, the HF sound level was higher or at least as high as the sound level of the corresponding LF component. When recorded from the front, the HF level was always considerably higher than the LF level.

B. Frequency of occurrence of LF and HF components in the recording sample

(a) The LF component was to be observed in all click trains recorded from the dorsal, frontal, lateral and ventral positions. Ventrally the LF signal was weaker but still measurable at the base of the flippers. Weak LF signals could also be detected from a short distance in the small tank when the whales had turned away from the hydrophone.

(b) The HF component was present in all click trains recorded in the dorsal and frontal position, in 1 out of 10 click trains recorded in the lateral position and in 4 out of 21 click trains recorded in the ventral position.

C. Reconstruction of the sonar field on the basis of frequency and intensity patterns

As soon as the whale uttered a sound, the LF component was clearly detectable from the entire front part of the body. It was therefore much less directional than the simultaneously produced HF component which could only be measured from a limited area of the front part of the body and declined rapidly in intensity at the edge of this area. Differences in the propagation characteristics of high and low frequencies are due to different degrees of diffraction (cf. the Huyghen's principle; Purves, 1966). But still the LF component was to a certain extent directional. It was more intensive in the *ventro-lateral than in the dorso-frontal area*. It is significant that compared with the HF component the LF component was particularly weak with the hydrophone in the front position where the HF component was strongest. Caudalwards the LF signal could still be detected at the base of the flippers whereas the HF component was completely lacking. Thus the axis of the LF sound field seems to be shifted downwards from the axis of the rostrum.

This comparison of signal intensities at different recording positions can only provide a qualitative estimate of the sonar field since it was not known with sufficient accuracy (a) how far the whale was from the hydrophone when the sound was uttered, and (b) whether intensity was always constant at the signal source. Nevertheless, comparison between signals from a number of recordings from the same and from different positions revealed a clear tendency for uniform signals to be recorded from the same position. This applies to echolocation signals emitted when the whales were swimming towards prey.

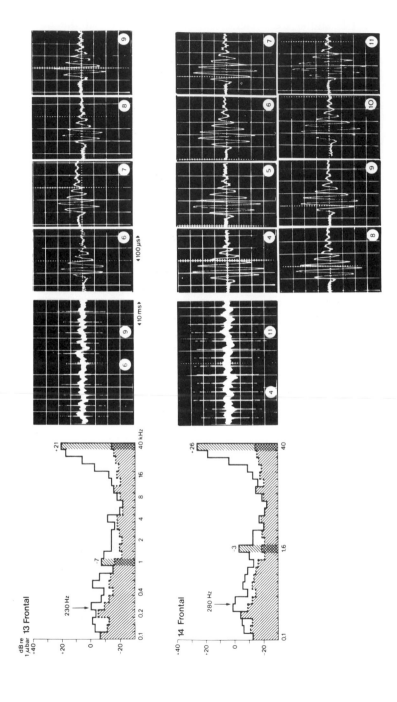

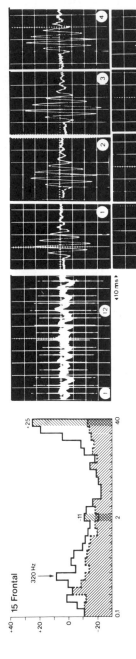

Fig. 7I (A–B) Sonar signals with animal swimming directly towards the hydrophone. In analysis No. 13 the animal was about 50 cm from the hydrophone; in the following Nos 14 and 15 it was less than 50 cm away. As the animal approached, the repetition rate of the clicks rose from 230 to 320 Hz, and the dominant frequency of the LF component increased from 1 kHz to 2 kHz. Note the regularity of the consecutive HF clicks. The clicks in this position have a simple structure and show no hangovers. Details of HF clicks 6–9 are shown from the oscillogram corresponding to analysis 13, clicks 4–11 are shown from analysis 14 and clicks 1–12 from analysis 15.

D. Signal spectrum and click shape as a function of repetition rate and recording position

Dominant frequencies in the LF and HF range:

The dominant frequency of the LF component appeared to be only slightly dependent on the position of the hydrophone, and ranged from 0·5–2 kHz. From the dorsal, lateral and ventral position, frequencies of 1·25–1·6 kHz were observed. In the front part of the field, the dominant frequency rose, as the whale approached the hydrophone, from 1 kHz or less to a maximum of 2 kHz (Fig. 71A, B). The fluctuations in the dominant frequency could not in all cases be attributed to differences in position. Other factors affecting it were the repetition rate and possible active changes in the individual signals emitted by the whale.

The HF component frequency was always highest in the central part of the field (frontal and rostro-ventral area), and lower at the edge of the HF beam (rostro-dorsal area). In other areas the HF component was lacking although the LF component was almost unchanged. When an HF component was to be observed, it was dominating the LF component in the 1/3 octave spectrum by an average of 15 dB. In the front part of the field the HF component was sometimes as much as 35 dB higher in intensity than the LF component. In the dorsal field, however, the HF component was only about 10 dB higher than the LF component.

Repetition rate as a component of the spectrum:

Figures 70 and 71 show that the repetition rate of the click train has a major impact on the LF component spectrum pattern. When a strong HF component was present, the repetition rate and its harmonics were dominant over the individual LF pulses. This was mainly the case in the frontal recording position.

At high repetition rates the individual LF pulses merged into a continuous signal. The resulting interferences showed up in the spectrum in the form of discontinuities and secondary dominant frequencies.

The effects described in subparagraphs one and two mean that changes in the LF component are difficult to interpret. They might either be merely the result of a change in repetition rate, or else an indication of an altered dominant frequency in the individual clicks.

HF pulse shape in different sectors of the sonar field:

In the central (front) sector of the emission field intensive clicks, approximately 175 μs long, were to be observed. As long as the whale swam straight towards the hydrophone, without turning its head sideways, successive clicks remained the same length and contained a constant number of about five periods (Fig. 71).

In the peripheral sectors of the acoustic field, weaker but similarly shaped clicks were to be detected, but accompanied by more or less marked hangovers. This meant that total pulse length in these sectors rose to 250–350 μs. The hangovers were least evident in the ventral sector where the signals came closest to frontal signals in intensity and pulse form. Pronounced hangovers could be recognized in the frequency spectrum by a flatter rise in the HF component in the frequency range between 4 and 25 kHz (cf. Fig. 70A/3).

Figure 70 shows two cases of secondary pulses preceding and following the main pulse at an interval of 400 or 500 μs. Both examples are probably echoes from the nearby prey or the nearest pool wall, one the echo of a preceding pulse and the other an echo of the pulse shown on the photograph. In neither case did the secondary pulses reach the same intensity as the main pulse.

Ventral sound propagation and pterygoschisis:

In an earlier work Pilleri (1979) drew attention to the significance for the ventral sonar field of incomplete ventral screening as a result of diverging pterygoid sinuses in a number of cetaceans. The two pterygoids with their extended hamuli provide a support for the palatopharyngeal muscle which encircles the larynx in the form of a sphincter and is prolonged into the nasopharynx. In the majority of delphinids the pterygoid bones cling closely to the vomer, while their hamuli meet in the middle and are entirely pneumatized. The result is that they provide complete ventral screening for the clicks produced in the larynx. This means that the sounds can only propagate in the direction of the vomer and upper jaw, in the form of a rostral field. In the white whale, however, the median edges of the two hamuli diverge (Fig. 63B) to form a rostrally open angle of about 64° (pterygoschisis). Pneumatization is confined to the lateral half of each hamulus. The area between the two hamuli constitutes the soft palate and is permeable to sound. The pars externa of the palatopharyngeal sphincter continues posteriorly as the thyropharyngeus and inserts on the mesial aspect of the thyroid cartilage, consequently the low-frequency vibrations of this cartilage escape through the soft palate together with high-frequency vibrations from the tip of the glottis, and at the same repetition frequency.

V. Development of a Sound Propagation Model

A simple sound propagation model can be built up on the basis of the empirically determined sonar field (Pilleri *et al.*, 1976, 1979) and the structure of the whale or dolphin skull. The model permits qualitative predictions concerning the shape and spread of the emission field. Measurements of the skull and air-sac system of the head provide the initial data for the model. As discussed in earlier papers (Pilleri *et al.*, 1976, 1979; Purves and Pilleri, 1973) the entire accessory air-sac system of the skull has a sound-screening function. Let us first consider the example of *Inia*. The pterygoid sinuses are integrated with the bones and form, together with the epicranial sacs, and in particular the premaxillary sacs, a sound-proof cavity around the sound source. A cross-section made through the dolphin's head at the base of the rostrum thus reveals a plane surrounded by air-filled systems which includes the cross-section of the upper jaw, and in *Platanista* the sectional area of the maxillary crests as well. This cross-sectional area may be regarded as the effective sound-propagating surface and its directivity may be calculated by the help of the model. For this purpose they assume that the sound is radiated regularly from the entire surface area. As a further simplification they assume that the area is roughly circular in shape. Thus with a given radius a and a given frequency f, the far-field directional characteristics may be calculated as follows:

$$D = 20 \log \frac{2 \, \mathcal{J}_1(ka \sin \theta)}{ka \sin \theta} \text{ [dB]}$$

where $\mathcal{J}_1$ is the Bessel function of the first order, $k = f. 2\pi/c$, the wave number, f is the frequency in Hz, $c = 1550$ m/s is the speed of sound in sea water, a is the radius of the source in m, and θ is the directional angle from the axial direction of the source. D is the directivity factor of a rigid, vibrating piston with circular cross-section mounted in an infinite baffle.

A polar plot of computer-calculated values of D shows the relative sound pressure levels at a constant distance from the sound source, but at varying angles to the axis of the source. This two-dimensional figure can be interpreted as a plane section through the acoustic field of the source in question. It shows a principal maximum in the axial direction, and on both sides of it usually one or more minima and secondary maxima of lower sound pressure. If we take the assumed horizontal and vertical diameter of the sound source as a basis for calculation, the resulting graphs can be directly compared with the acoustic fields measured for the whale or dolphin in the vertical and horizontal plane. Figure 72A, B shows a

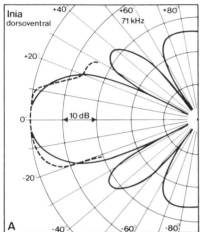

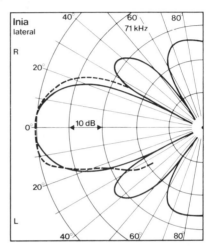

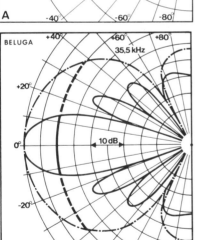

Fig. 72 (A–B) Comparison of theoretical directional characteristics with measured emission fields of *Inia* (A) and estimate of the emission field of *Delphinapterus* (B). A: unbroken curves: directivity of a rigid, vibrating piston 5·5 cm in diameter at a frequency of 71 kHz. Broken curves: directional characteristics for sonar signals measured from the animal in the dorso-ventral and in the lateral plane of inter-section. Measuring bandwidth was 1/3 octave with a centre frequency of 80 kHz and band-edge frequencies of 71 and 89 kHz. B: Unbroken curve: directivity of a rigid vibrating piston 16 cm in diameter. Model representation of the horizontal (lateral) extension of the sonar field. Broken curve: directivity of a 5·5 cm diameter piston. Model representation of the dorso-ventral extension of the sonar field. The calculation is based on a frequency of 35·5 kHz. The unbroken and broken arc segments indicate the −10 dB beam widths of 26° in the horizontal plane and 78° in the vertical plane.

superimposition of the directional characteristics as measured from the animal and calculated from the model in the case of *Inia geoffrensis* as well as estimated directivity for *Delphinapterus leucas*.

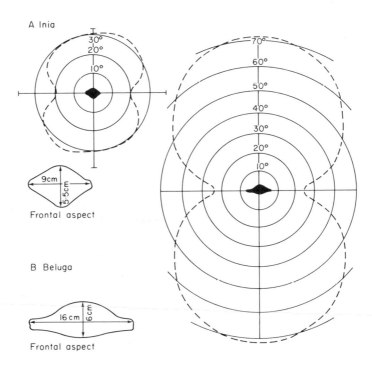

Figs 73 (A–B) Model calculation: lines of equivalent sound pressure level in the far field, looking from the front along the acoustic field axis. The graphs show the − 10 dB width of the main lobe in cross-section. The angular notation refers to the diameter of the concentric circles. The calculation was based on the measured cross-sections at the rostrum base shown in black in the centre of the graphs, and in enlarged form on the left. The horizontal and vertical bars correspond with the − 10 dB beam width of the sonar fields measured from the animal at approximately the same frequencies, and are shown by way of comparison. A: *Inia*: Frontal view (o°). In the model the beam width (− 10 dB) at 80 kHz is 20° in the horizontal direction, and rather more than 30° in the vertical direction. B: *Delphinapterus*: Frontal view (o°). The beam width (− 10 dB) at 40 kHz is about 25° in the horizontal direction, and more than 70° in the vertical direction. Ventral extensions of the sonar field due to the pterygoschisis are not shown in the graph.

A. *Inia geoffrensis*

In an adult *Inia* the average diameter of the cross-sectional area at the base of the rostrum is 5–6 cm. The horizontal and vertical dimensions of the cross-section are almost identical. We should therefore expect an emission field with a similar width in the dorso-ventral and horizontal plane. The acoustic field predicted by the model of a circular, sound-propagating rigid piston 5·5 cm in diameter agrees well with the field actually measured from the dolphin. This is true not only of the field width, but also of the decrease in sound level at the edge of the central maximum (Figs 72, 73A).

B. *Delphinapterus leucas*

In the case of the white whale the model can be used to make predictions about the probable shape of the emission field which correspond well with the qualitative experimental findings. The upper jaw of *Delphinapterus* is much wider (16 cm) than it is high (6 cm). This is an important difference with respect to *Inia* whose upper jaw is almost circular in cross-section. But just as in *Inia*, the broad sectional area is limited dorsally by the superimposed premaxillary sac. Ventrally, however, it is not completely covered by the pterygoid sinus. The two, only partly pneumatized, pterygoid hamuli diverge widely in the middle (pterygoschisis). Figures 72B and 73B show the hypothetical emission field of the white whale as calculated from the model, based on the cross-section of the upper jaw, with a HF signal frequency of 35·5 or 40 kHz respectively. The model of the white whale's emission field is characterized by a broad dorso-ventral and a small lateral extension. The results obtained from recordings made in the aquarium agree well with the theory. For 1·6 kHz centre frequency of the LF component, almost non-directive radiation is to be expected on the basis of the measured diameter of the sound source.

VI. Sonar Field in *Neophocaena phocaenoides*

Having thus established that the white whale, *Delphinapterus leucas*, has a low-frequency as well as a high-frequency component of the sonar emission, and that the low-frequency pulse emanates mainly from the throat area whilst in the frontal area the high-frequency component is dominant, the time has come to examine the sonar records of *Neophocaena phocaenoides* taken in Kai Creek in the Indus Delta, since this species too has a marked pterygoschisis.

The porpoise sounds were recorded with a Nagra-IV L tape recorder in conjunction with a conditioning amplifier and a Bruel and Kjaer 8100 measuring hydrophone with a nominal sensitivity of -110 dB re 1 V/μbar. The frequency response was limited by the tape recorder to the 30 Hz–35 kHz (-3 dB) range at 38 cm/s. The original tapes were used when analysing the signals in the laboratory. A 564 B Tektronix oscilloscope was used to represent the signal in the time domain. It was decided not to carry out any frequency analysis because of the low signal-to-noise ratio of some of the recordings.

A. Model calculation

A second step of the research was to develop a simple model of sound radiation, as already described for the white whale (Zbinden *et al.*, 1980). The model permits estimates of the extent and direction of the sonar field on an anatomical basis. The estimates of directivity were based on two sound-propagating structures with partly overlapping radiation areas. The first was the sectional area at the base of the upper jaw and the second was the gap between the two pterygoid hamuli, or sinus pterygoidei (pterygoschisis; Figs 74 and 75).

(a) Model based on the cross-section of the rostrum base:

This model representation is suitable for an approximate estimate of the frontal emission field.

The cross-section of the rostrum base is considered as a vibrating, rigid, circular piston. The directivity of the source is calculated as follows:

$$D = 20 \log \frac{2 \, \mathcal{J}_1(ka \sin \theta)}{ka \sin \theta} \; [\text{dB}]$$

where D is the directivity in dB, $\mathcal{J}_1$ the Bessel function of the first order, $k = 2\pi f/c$ the wave number, f is the frequency in Hz, $c = 1550$ m/s, the speed of sound in sea water, a is the radius of the source in m and θ is the directional angle from the axial direction of the source.

The graph of D as a function of the angle θ appears as a curve of the relative sound pressure level at a constant distance from the sound source,

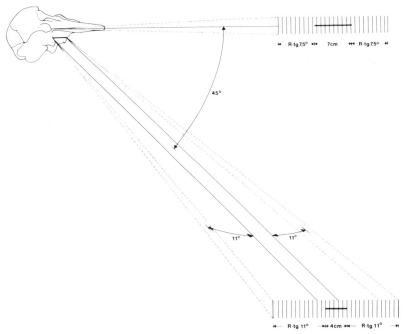

Fig. 74 Three-dimensional diagram of estimated −10 dB width of the frontal and ventral sonar fields of *Neophocaena phocaenoides* and *Phocoena phocoena* at a frequency of 140 kHz. Frontally the horizontal angle of propagation is 15° and ventrally 22°. The rostrum base width of 7 cm and the pterygoschisis of 4 cm are indicated with horizontal bars. At a distance of $R = 50$ cm from the head of the porpoise, the frontal field width ($A = 2$ Rtg 7·5° + 7 cm) is a total of 20 cm and the ventral field width 23·5 cm.

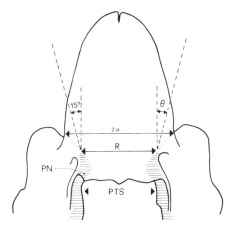

Fig. 75 Schematic ventral view of the skull of *Neophocaena phocaenoides*. PN = sinus pterygoidei (hatched) PTS = pterygoschisis. R = width of schisis slit. 2a = width of rostrum base, θ = horizontal angle of propagation of sound waves radiated in the ventral direction from the pterygoschisis.

but at different angles to the source axis. The values are represented two-dimensionally to produce two diagrams of the $-90°$ to $+90°$ range, one based on the horizontal and the other on the vertical source diameter. The two curves thus represent the horizontal and the vertical plane of section through the sonar field. The width of the principal maximum in each case can be read directly from the polar diagrams.

Beam widths are usually given as -3 dB, -6 dB or -10 dB data or as the angle between the first zero points of the directivity function on both sides of the central maximum, the angle of the latter being easier to calculate as follows (minimum condition with a circular sound source):

$$\sin \theta = \pm \frac{mk\lambda}{a}$$

where a is the radius in m, λ is the wavelength in m and for the first order minimum ($M = 1$), the factor $k = 0.61$.

(b) Model based on the gap between the pterygoid sinuses (pterygoschisis):

The second model is suitable for approximate calculation of the lateral field width in the ventral sector of the sonar field, where the pterygoschisis is responsible for sound radiation.

The pterygoschisis is regarded as a long, sound-radiating slit or line source. The directivity of a lone source is calculated as follows (Clay and Medwin, 1977):

$$D = 20 \log \left| \frac{\sin \left(k \dfrac{R}{2} \sin \theta \right)}{k \dfrac{R}{2} \sin \theta} \right| \quad \text{dB}$$

where k is the wave number, R is the width of the slit and θ the directional angle from the axial direction of the source. Sin θ is calculated in the degree modus, whereas the sine of the expression $(k\dfrac{R}{2} \sin\theta)$ has to be calculated in radian mode. Just as in the case of the circular piston, the function is periodic. The minima are calculated according to the formula:

$$\sin \theta = \pm \frac{n\lambda}{R}$$

with R as the width of the slit and λ the wavelength in m. For the first order minimum, $n = 1$.

In order to compare the results for the position of the first order minima as calculated according to method (a) and method (b), we let $R = 2a$, so that:

$$\sin \theta = \frac{0 \cdot 61 \lambda}{a} = 1 \cdot 22 \frac{\lambda}{2a}$$

Thus with an identical source diameter, the field calculated according to the circular piston formula is wider by a factor of $1 \cdot 22$ than the field obtained with the slit formula. This factor reflects the impact of the shape of the sound source on directional radiation. The dimensions being the same, a source in the form of an extended slit at right angles to its longer dimension gives better directivity than a source in the shape of a circular piston with a diameter equal to the width of the slit. In the case we are discussing, the result is that ventrally, in the area governed by the pterygoid slit, the lateral width of the sonar field is roughly the same size as the width of the frontal field, although the lateral dimensions of the slit are smaller than those of the cross-section of the rostrum (cf. Fig. 74).

In addition to the beam widths from minimum to minimum, the -10 dB beam widths were also estimated with both methods. The results are shown in Table 9.

VII. Sounds of *Neophocaena phocaenoides*

The recordings of the sounds made by *Neophocaena phocaenoides* in the course of several days in Kudi Creek and Khai Creek revealed only a single type of sound in the frequency range of 30 Hz to 35 kHz.

Very weak, low-frequency clicks were recorded, repeated at varying rates. The absolute sound pressure level of the signals lay between $+5$ and $+8$ dB re 1 μbar at 1 m. With the apparatus we were using in those days which had a noise level of -8 dB re 1 μbar, the sounds could only be recorded when the porpoises were very close to the hydrophone. For example, it was not possible to record the clicks make by a group of six *Neophocaena* swimming past the stationary boat at a distance of only 6 m. This was because, with a source strength of 5–8 dB, the signal level reaching the hydrophone at a distance of 6 m was no more than -11 to -7 dB, which was below the noise level of our apparatus. When the attempts made to record the sounds of free swimming porpoises proved unsuccessful, it was decided to catch the animals with nets and keep them in a closed-off section of the creek so that we could investigate them from nearer at hand.

The shape and frequency of the clicks of the three porpoises kept in semi-captivity varied as a function of the position of the recording hydrophone. First described are the clicks recorded from a frontal position at a distance of about 1 m from the animal. Click length fluctuated between 1·3 and 3·4 ms, with a period number of from 2·5 to 6·5. The frequency of the most intensive period was in all cases about 2 kHz (ranging from 1·6–2·2 kHz). The first 2·5 periods were always the most intensive (Fig. 76). In signals which were more than 2·5 periods long, the following third, fourth, fifth and sixth periods consistently showed a roughly exponential decrease in amplitude. The data for click length and period number fluctuate considerably owing to the low signal-to-noise ratio and represent only approximate values.

The clicks were always produced in series, with a repetition rate in the range of 20–150 Hz (Fig. 77). The average rate (±) of 347 series was 60·8 clicks/s (s = 18·4). With the hydrophone in the frontal position, short peaks of high level often appeared at the beginning of the clicks (Fig. 76). It was subsequently discovered by means of experiments that these were artefacts caused by overload of the tape recorder by high-frequency (HF) signal components lying outside the frequency range of the recorder. Overload of this kind presupposed an HF component with a level 35 dB higher than that of the low-frequency (LF) component. This implied an absolute sound pressure level for the HF component of not less than +40 dB re 1 μbar. The position of the artefact in the LF click tells us something about the relative position and approximate length of the original HF signal component. As in the case of *Delphinapterus leucas* (Zbinden *et al.*, 1980) we noted an exact synchronization between the LF and the HF clicks. The HF artefact was always superimposed on the first, usually positive half-period of the LF click. Its length pointed to an HF click of about 100 μs. The HF artefact never occurred in isolation, but always in association with the LF click. So far we have not been able to draw any conclusions about the pulse shape or frequency of the HF component. Clearly, however, the sonar signals of *Neophocaena phocaenoides* consist of two synchronized components with frequency ranges, one in the audible range and the other in the ultrasound range, lying very far apart.

In the case of moving animals LF clicks with HF artefacts were much less often observed than LF clicks without them. This is due to the highly directional radiation of the HF component. Where an artefact was present

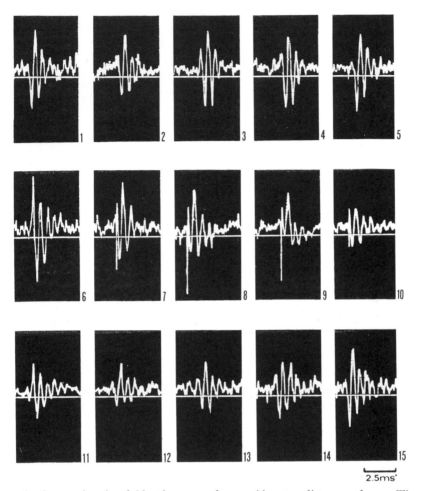

2.5ms

Fig. 76 Sonar signals of *Neophocaena phocaenoides* at a distance of 1 m. The 15 successive clicks show a first increasing and then decreasing HF component, as the porpoise turns towards and then away from the hydrophone. Clicks 7–9 represent an approximately frontal hydrophone position.

the HF component of the signal must undoubtedly have had a considerably higher level than the LF component. At the same time absence of the artefact did not necessarily imply that the signal contained no HF component, but only that its level could not have been substantially higher than that of the LF component.

The recordings described permit an exact determination of the sound level distribution of the LF component at different recording positions

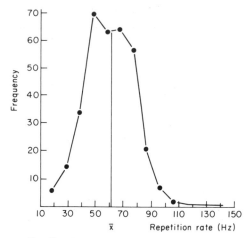

Fig. 77   Frequency distribution of the repetition rate of a *Neophocaena phocaenoides* in Kudi Creek, Indus Delta.

(Table 8). In the case of the HF component they can only establish the direction of the strongest radiation, as the artefacts it gave rise to in the LF range only showed up when the tape recorder was severely overloaded.

A. Near-field characteristics of clicks recorded at different transducer positions

Figure 78 shows a diagram of the profile of the head of *Neophocaena phocaenoides*. Sounds recorded from different positions are illustrated by bar diagrams and the shapes of the corresponding sonar clicks.

B. Nostril position (not shown in Fig. 78)

Broad-band pulses with low-frequency (about 500 Hz) and high level (+22 dB re 1 μbar at the nostril). The pulses were strongly frequency-modulated and had three periods with a total length of about 7 ms (Fig. 79).

C. Melon (dorsal), rostrum tip (frontal), throat (ventral), and corner of the mouth (lateral) positions

The clicks were less broad-banded and less strongly frequency-modulated than signals recorded with the hydrophone close to the nostril. The LF

Table 8 Click properties and pterygoschisis

		Phocoena phocoena	*Neophocaena phocaenoides*	*Delphinapterus leucas*
Form of clicks, frontal				
Position of HF component		during first period of LF pulse *1, *3	during first half-period of LF pulse	during first third of LF pulse
Pulse length	LF	0.5–5 ms *1, 3 ms *2	1.2–3.4 ms	2–4 ms
	HF	100 μs *3	?	75–250 μs
Broad-band sound pressure level in dB at 1 m, frontal	LF	−7 to +8 dB (0 dB) *3	+4 to +8 dB	−5 to +5 dB
	HF	+32 to +49 dB *3	?	(+20 dB to +25 dB) *4
Frequency in kHz	LF	2	2	1.2–1.6
	HF	110–150	?	40 *4, (40/80/120) *5
Number of periods	LF	2–4 *1, *2	2.5–6.5	2–3
	HF	7–11 *3	?	(3–5) *4
Functional width of schisis		4 cm $\triangleq$ 57 % of the rostrum base	4 cm $\triangleq$ 57 % of the rostrum base	5 cm $\triangleq$ 31 % of the rostrum base
Ventral field	LF	detected	detected	detected
	HF	detected	predicted	detected *4

*1 SCHEVILL *et al.*, 1969,   *2 DUBROVSKII *et al.*, 1971,   *3 MØHL and ANDERSEN, 1973,   *4 Frequency range limited to 40 kHz, ZBINDEN *et al.*, 1980,   *5 GUREVICH and EVANS, 1976.
Fluctuations in pulse length and number of periods are due at least in part to the limited accuracy of the measurements.

pulses had a frequency of about 2 kHz (range 1·6 2·1 kHz) and showed a more or less marked exponential decay. The total length varied in a range from about 2 ms to 7 ms, of which 1.5 ms to a maximum of 2·5 ms had high intensity at highest frequency. The number of periods recorded from all positions was between 2·5 and 7·5, of which there were 2·5 to 4 periods of high amplitude. Figure 78 shows the typical shapes of signals recorded from the different positions. It is not, however, possible to

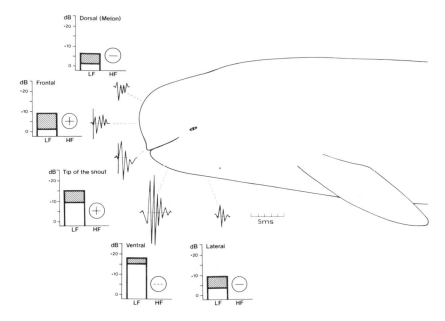

Fig. 78 Diagram of near-field properties of the sonar clicks of *Neophocaena phocaenoides* recorded in different positions around the head of the porpoise. Each position is illustrated by a schematic oscillogram of a typical click, as well as the absolute broad-band sound level of the low-frequency (LF) component. The hatched region of the bars indicates the range of fluctuation between the measured values. A+ sign indicates a high-frequency (HF) component in the recording. A — sign shows that there was no HF component.

indicate accurate quantitative differences in frequency, click length and period number because the rather low signal-to-noise ratio meant that details were lost and the pulse limits could not be exactly determined. This also explains the considerable range of fluctuation in the data quoted for pulse length and period number.

Nevertheless, clicks recorded from the dorsal, frontal, ventral and lateral position were clearly distinguishable from those recorded from the nostril (Fig. 79), since they had an average length of 3·5 ms and an average frequency of 2 kHz, whereas the nostril clicks were an average of 7·2 ms long and had an average frequency of 500 Hz. *This seems to indicate that signals recorded at the nostril were passive in character, having a longer rise and decay time and a much lower frequency. They were probably produced by vibration of the soft structures in the upper nasal tract.*

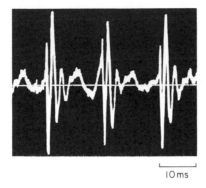

10 ms

Fig. 79 Oscillogram of signal recorded at the nostril.

Apart from this, there were definite differences in the sound level of signals recorded from different positions. The maximal near-field sound pressure level occurred ventrally (at the throat) with an average maximum of +16 dB re 1 μbar which generally declined in the dorsal direction. At the rostrum tip the sound pressure level was only about two-thirds as high at +11 dB, while frontally and dorsally (45°) it was a good two-thirds lower (+5 dB and +4.5 dB respectively). In the lateral position at the angle of the mouth, the sound pressure level was an average of +6 dB re 1 μbar. All sound pressure levels were measured peak-to-peak.

HF artefacts superimposed on LF signals occurred with the hydrophone in the dorso-frontal position and at the tip of the rostrum.

This means that the HF signal component is concentrated into a narrower beam than the LF component. After the experience we would expect to find an HF component in the ventral region as well. This, however, was not the case in the recordings made from the ventral position in 1979. This is attributed to the fact that the hydrophone was not positioned with sufficient accuracy. With the narrow ventral field expected, even small deviations in the position of the hydrophone on the skin of the throat could lead to disappearance of the strongly directional HF component.

The model predictions for the directional characteristics of the HF field are based on an HF component frequency of 140 kHz. According to Møhl and Andersen (1973) this is the frequency of the HF component in the signals of *Phocoena phocoena*. Although the frequency of the HF component in the *Neophocaena* signals is still not known, the observations indicate that it must be very high, so that it seems reasonable to postulate the same frequency as that produced by the related *Phocoena*. The assumption seems particularly justified in view of the very similar morphology of the structures of the head responsible for sound production and radiation in the two porpoises.

Figure 80 shows estimated directional characteristics of *Neophocaena*, in the horizontal (lateral) and vertical (dorso-ventral) plane. At 15° the horizontal —10 dB width of the radiated field is only just half the vertical width, so that on the basis of the rostrum cross-section alone we expect a field which is considerably higher than it is wide. Figure 81 shows the estimated field cross-section in frontal view. The 4 cm wide pterygoschisis means that account must be taken of an additional extension of the sonar field in the ventral direction. The relation between the horizontal beam width in frontal and in ventral (—45°) position is illustrated by Fig. 74. The figure indicates that at a distance of 50 cm from the sound source, the extended ventral field produced by the pterygoschisis is only slightly broader than the frontal field emanating from the cross-section of the rostrum. This is true despite the smaller source diameter R for the ventral field (Fig. 75). The effect can be explained by the higher directivity of a line source compared with a circular piston source. The angle of a total of 30° (±15°) shown in Fig. 75 indicates the minimum-to-minimum beam width of the schisis signal. The corresponding —10 dB beam width is 22° (±11°; *cf.* Table 9).

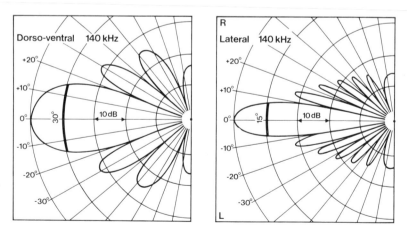

Fig. 80 Estimated directional characteristics of the emission field at a frequency of 140 kHz in the vertical (dorso-ventral) and the horizontal (lateral) sectional plane. The estimate is based on a rostrum base height of 3·5 cm and a width of 7 cm. The thick unbroken sections of arc show the —10 dB beam width.

Table 9:   Estimated total beam width in the horizontal plane.

Species		−3 dB	−10 dB	Angle of the minimum	Base (cm)	Frequency
Neophocaena phocaenoides and *Phocoena phocoena*	Frontal	9°	15°	21°	7	140 kHz
	Ventral (schisis)	13°	22°	30°	4	
Delphinapterus leucas	Frontal	15°	26°	36·5°	16	35·5 kHz
	Ventral (schisis)	43°	74°	100°	5	
Inia geoffrensis	Frontal	22°	37°	52·5°	5·5	71 kHz

VIII. Sonar System in *Phocoena phocoena*

The echolocation sounds produced by the common porpoise have been well described in the literature, and consist of both narrow-band, high-frequency clicks (up to 100 kHz, according to Dubrovskii *et al.*, 1971; or 110–150 kHz, 40 dB re 1 μbar at 1 m, according to Møhl and Andersen, 1973) and synchronous low-frequency components (2 kHz, 1 μbar at 1 m: *idem*; Schevill *et al.*, 1969). It is also known that the HF component is very strongly directional (Møhl and Anderson, 1973).

A tape recording of sounds made by *Phocoena phocoena* in captivity was lent by Dr Andersen of Odense. During recording in the tank the sounds had been filtered with a 2 kHz (1 octave) band-pass filter. The recordings were made with the hydrophone at a distance of 50 cm from the porpoise, and positioned once frontally (0°) and twice ventrally (−35° and −50°). Several click series were recorded from each position. We determined the absolute sound level of the click trains, the position of the HF-induced spikes within the individual low-frequency clicks, and the direction of the axis of the HF-sonar field. The recordings confirmed the level data for the LF component quoted above. At 2 kHz we determined an average

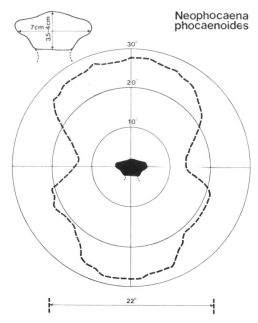

Fig. 81 Estimated cross-section of frontal emission field of *Neophocaena phocaenoides* at a signal level of −10 dB and a frequency of 140 kHz. The estimate was based on the cross-section of the rostrum base shown in the centre of the figure and above left.
The pterygoschisis is indicated by a broken line as a ventral extension of the rostrum cross section. In the lower part of the figure the 22° angle marks the −10 dB width of the field radiated from the schisis in the ventral direction. The total extension of the frontal field in the vertical direction, not counting the ventral field, is almost 30°, while the extension in the horizontal direction is about 15°.

absolute sound level of −3 dB re 1 μbar (1-octave filter, hydrophone in the frontal position, at a distance of 0·5 m from the tip of the rostrum). The average LF sound level in the ventral position was identical (Fig. 82). The fact that the 40 dB higher HF component, as in the recordings of the *Neophocaena* signals, resulted in an artefact superimposed on the LF click, means that it was possible to estimate the axis of the HF-beam and the approximate relative position of the HF click within the LF pulse.

The position of the HF artefact shows that the HF click must be super-imposed on the first period of the LF pulse. The study by Schevill *et al.* (1969) indicates that this is in fact the case.

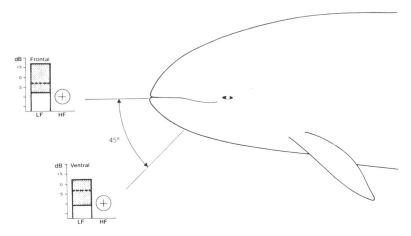

Fig. 82 Diagram of some properties of the sonar signal of *Phocoena phocoena* recorded in the frontal and ventral position. The hatched area of the bars indicates the range of fluctuation and the broken line shows the average value of the broad-band sound pressure level of the low-frequency (LF) sound component. An HF component was recorded in both positions.

Overload of the recording equipment caused by the HF signal occurred with the hydrophone in both the frontal and ventral position (to at least 50° ventral of the rostrum axis). This means that the HF field in *Phocoena* extends from the frontal to far down in the ventral direction. Figure 74 is thus approximately valid for the emission field of *Phocoena* as well. The fact that the cross-section of the rostrum base is almost identical in shape and size in the two species means that they both produce fields with the same widths (Table 9 and Fig. 80) and very similar cross-sections (Fig. 83).

The findings show that not only the sonar signals but also the structures responsible for sound production and transmission are very similar in *Neophocaena phocaenoides* and *Phocoena phocoena*. Both species produce echolocation clicks with a synchronous HF and LF component. The two LF components are indistinguishable in terms of frequency and number of periods. The source sound levels of the LF and HF components are of the same order of magnitude. In both cases the HF component is detected synchronously with the LF signal. The HF click is always superimposed on the first period of the LF signal and has a length of at most 100 μs.

The HF components in both cases are strongly directional. The HF emission field is prolonged in the ventral direction. This means that the sounds are radiated not only rostro-dorsally and rostrally, but they can

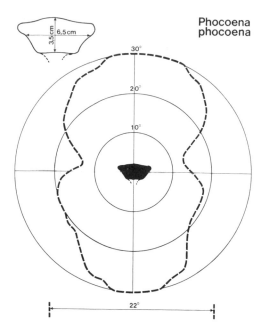

Fig. 83 Estimated cross-section of the frontal emission field of *Phocoena phocoena* at a signal level of −10 dB and a frequency of 140 kHz. Other explanations as in Fig. 81.

also be recorded in the region of the throat. This has been clearly demonstrated in *Phocoena*. In *Neophocaena* it was not yet possible to detect the probably narrow ventral field with the simple methods used at the time of recording. The very great morphological similarity between the sound-radiating structures in the two species makes it extremely likely that *Neophocaena* also possesses a ventrally extended emission field. The almost identical properties of the sonar sounds go hand in hand with the matching morphological structure of the sound conducting and producing organs. The cross-section of the rostrum base is in both cases about 7 cm wide and the width of the pterygoschisis 4 cm (Figs 75 and 84). The diameters of the two sources result in an angle of propagation of 15° in the lateral direction for the frontal field, and a not much larger angle of 22° (−10 dB beam width) for the width of the ventral field. The fact that a slit has better directivity by a factor of 1·22 than a circular source of the same size, means that in spite of the small width of the slit, the ventral field is not much wider than the frontal field. The result is a sonar field with a typical high, narrow form.

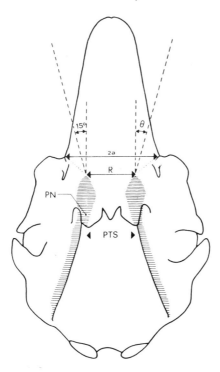

Fig. 84   Schematic ventral view of the skull of *Phocoena phocoena*. For
abbreviations see Fig. 75.

During the studies of the sonar field of *Neophocaena phocaenoides* the
highest LF sound level was definitely noted in the ventral part of the field.
Figure 78 shows the sound level measurements in the near field. This
fact provides important evidence of a laryngeal origin for the sounds.

If, as a number of authors still believe, the sounds were really produced
epicranially in the region of the middle pair of sacs of the nasal canal, and
if the melon in fact acted as an "acoustic lens", then the sounds ought to
be much louder with the hydrophone in the dorsal position close to the
melon. According to our measurements this is in no way the case. It was
in the region of the throat that the low-frequency sounds in the near field
always had a distinctly higher level than in other positions. When the
measurements were made at a greater distance from the porpoise, spherical
wave propagation of low frequencies meant that there was less difference
in the level at different recording positions (cf. the measurements for
Phocoena phocoena Fig. 82). A second anatomical argument can be adduced

for the origin of the sounds. Supposing the theory of an epicranial sound source were correct, the sounds could never be detected ventrally in the region of the throat, as they would be reflected in another direction by the premaxillary air-sac. *Only sounds produced in the larynx can radiate unhindered in the ventral direction, through the pterygoschisis.*

Lastly the question arises as to what is the physiological significance of the special form of sonar sounds and sonar field in species with an open pterygoid. It is not easy to provide an answer, as we know so little about the behaviour of these species and their use of the sonar system for locating and catching prey. In the case of *Delphinapterus leucas* Pilleri (1979) suggested that a ventrally extended sonar field would be useful for locating prey on the sea floor, and we know that the white whale includes flatfish in its diet, and at certain times of the year eats nothing else. No observations have so far been made on the way in which *Neophocaena phocaenoides* locates and captures shrimps. The fact that the sonar apparatus is the cetacean's most valuable sense organ suggests that in the course of evolution it has become ideally adapted for catching prey.

VII.   Phonation in the Blind Indus Dolphin *Platanista indi*

Anatomical and behavioural evidence in *Platanista* shows that it is the most highly specialized of all the Cetacea in echolocation and must be regarded as a unique example in this respect. The anatomical features of phonation and hearing are so intimately interdependent that some parts of the former have to be described simultaneously with those of the latter. The most striking feature is the development, at the lateral extremities of the frontal area of the skull, of a pair of large, pneumatized maxillary crests —the cristae maxillares (Fig. 85). These pneumatized bones are anterior extensions of the cavity of the middle ear. A small branch of the pharyngo-tympanic or Eustachian tube pierces the frontal aspect of the skull, becomes gradually wider and spreads into a fan-shaped array of elongated diverticula separated by bony trabeculae (Fig. 86). This system of air spaces is covered externally by a thin lamella of bone a few mm thick, but the interior wall is extremely thin, translucent and flexible, consisting of only a partially ossified periosteum. The cavities are lined by a mucous membrane and a ciliated epithelium which is continuous with that of the tympanic cavity. Figure 87 is a radiograph of the skull after injection, through the tympanic membrane, of the air sinus system with a radio-opaque, iodized oil, Lipiodol. In this preparation the brain had been removed so that the occipital area is missing but this did not affect the air sinus system. The radiograph shows that the area of the ear-bones, the tympanic cavity and the various extensions of the pterygoid sinus system constitute one continuous air-cavity.

When bony crests are present in most mammals they are usually provided for the attachment of muscles. It will be seen from Fig. 88 that no

Fig. 85 Lateral view of the skull of *Platanista indi*. CM = maxillary crest.

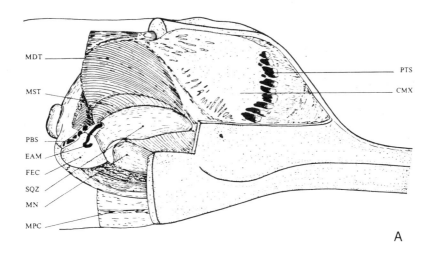

A

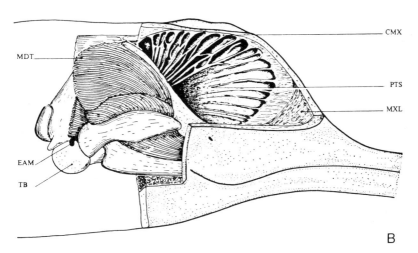

B

Fig. 86 (A–B)   A: Diagram of a dissection of the right side of the head of *Platanista indi* to show the external aspect of the maxillary crests and the anterior extremities of the maxillary air sinuses which lie immediately under the epidermis. B: Diagram of the same dissection after removal of the external osseous wall of the maxillary crests. The thin, unossified periosteal, mesial wall of the crests is shown with the radiating bony trabeculae which connect it to the external lamina.

CMX = maxillary crest, EAM = external auditory meatus, FEC = fibro-elastic capsule, MDT = deep temporal muscle, MN = mandible, MPC = panniculus carnosus muscle, MST = superficial temporal muscle, MXL = maxillolabialis muscle, PBS = peribullary sinus, PTS = pterygoid sinus, SQZ = zygomatic process of squamosal, TB = tympanic bulla.

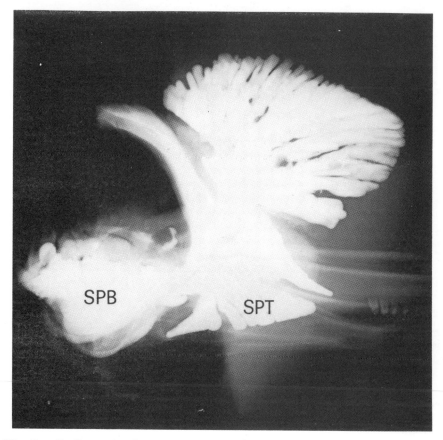

Fig. 87 Radiograph of a head of *Platanista indi* showing complete filling of tympanic cavity, pterygoid sinus and maxillary crests. SPB = peribullary sinus, SPT = pterygoid sinus.

muscles whatsoever are attached to the crista maxillaris, those operating the blowhole being attached to the skull below the level of the crests. Thus the latter can only have an acoustic function (see Chapter IV).

Platanista is peculiar in the sense that its anatomy consists of a mosaic of primitive and highly specialized characters. For instance, the ear bones are not as completely isolated from those of the skull by air spaces as in other odontocete cetaceans, but acoustic insulation of the sound-producing organ, the larynx, from the ears is achieved in quite a different manner.

Originating higher in the Eustachian tube there is another pair of air-sacs named the nasopharyngeal sinuses. These pass at first ventrally and

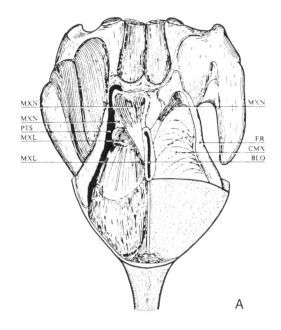

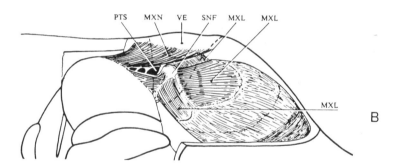

Fig. 88 (A–B) A: Diagram of the dorsal aspect of the head of *Platanista indi* to show the muscles of the blowhole and the bulbous mass of fibrous tissue which is the homologue of the "melon" of other delphinids. B: Diagram of the lateral aspect of the same dissection showing the ascending diverticulum of the nasofrontal sac. A small part of the posterior extremity of the maxillary crest and sinus is intercalated between the superficial muscles.

BLO = blowhole, CMX = maxillary crest, FR = frontal bone, MXL = maxillolabialis muscle, MXN = maxillonasalis muscle, PTS = pterygoid sinus, SNF = nasofrontal sac, VE = vestibule.

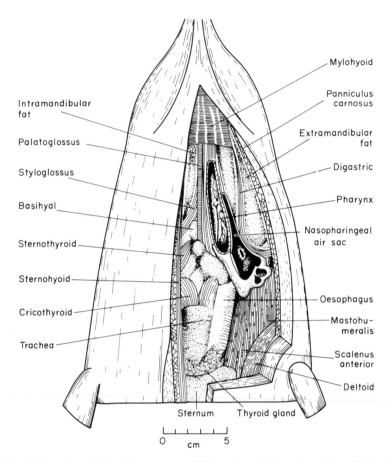

Fig. 89   Dissection of the throat area of *Platanista indi* to show the relationship between the nasopharyngeal air-sac and the upper, phonatory part of the larynx (see also Fig. 90).

then posteriorly to lie alongside the larynx (Fig. 89). In cross-section (Fig. 90: NPAS) they are seen to envelope almost completely the palato-pharyngeal sphincter and the glottis where the high-frequency sonar pulses are produced, therefore there must be almost complete acoustic isolation of the larynx from the ears. These sacs, first described but not figured by Anderson (1878), were likened to the gutteral pouches of the horse because the latter are also diverticula of the Eustachian tube. Whether there is any homologous significance in this similarity is unknown.

   There are no conspicuous laryngeal air-sacs in *Platanista* and the reservoir for containing recycled air, the premaxillary sacs, are relatively

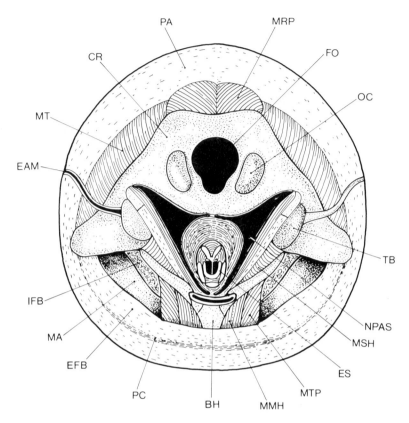

Fig. 90   Diagram of the posterior aspect of the skull of *Platanista indi* to show the relationship of the nasopharyngeal air-sac to the larynx and palatopharyngeal sphincter seen in cross-section.

BA = basihyal, CR = cranium, EAM = external auditory meatus, EFB = extramandibular fatty body, ES = epiglottic spout, FO = occipital foramen, IFB = inframandibular fatty body, MA = mandible, MMH = mylohyoid muscle, MRP = rectus posterior muscle, MSH = stylohyoid muscle, MT = temporal muscle, MTP = tensor palate muscle, NPAS = nasopharyngeal air-sac, OC = occipital condyle, PA = panniculus adiposus, PC = panniculus carnosus, TB = tympanic bulla.

small (Fig. 91: SPM). However, as the average diving time is only 30 s to 1 min such reservoirs are probably unnecessary.

The arrangements for pneumatic occlusion of the blowhole are present (Fig. 91: SNF) but lie at a considerable distance ventral to the entrance to the blowhole. There are no vestibular sacs as in most other odontocetes,

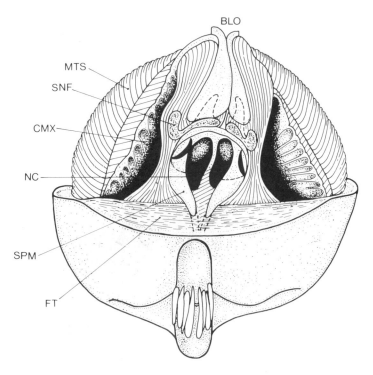

Fig. 91　Frontal view of a dissection of blowhole of *Platanista indi* after removal of the nasal plugs showing the relationship of the nasofrontal to the premaxillary sacs.
BLO = blowhole, CMX = maxillary crest, FT = fibrous tissue of the melon, MTS = superficial temporal muscle, NC = nasal canal, SNF = nasofrontal sac, SPM = premaxillary air-sac.

and this could be due to elongation of the vestibule of the blowhole as a result of the development of the maxillary crests. A similar situation occurs in the Bottle-nosed whale, *Hyperoodon*, which also has maxillary crests, although these are not pneumatized and are provided solely for the attachment of the massive blowhole muscles.

The palatopharyngeal recess into which the tip of the epiglottic spout protrudes is much more accurately sculptured to the configuration of the tip of the glottis than in other delphinids, and is deeply indented with mucous ducts (Fig. 92: RPP). A cross-section of the glottis in *Platanista* shows that it lacks the circular aperture between the cuneiform cartilages as seen in *Delphinus* and therefore according to our theory should not be

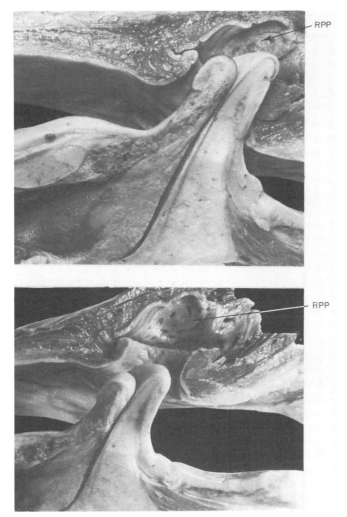

Fig. 92 Bisected larynx and posterior nares of *Platanista indi* to show the recessus into which the distal extremity of the epiglottic spout is intruded. Note the large entrances to mucous ducts in this area (c.f. also Fig. 30).

capable of producing whistles. The same situation applies to *Mesoplodon* (Fig. 93, A–C).

One of the points established in previous studies was that the sonar sounds emitted by *Platanista indi* originate exclusively from the larynx. The sonar signal consists of uniform clicks whose shape and frequency spectrum are not actively modified by the animal. Any changes in the shape and frequency content in the case of a swimming dolphin could be attributed to the directionality characteristics of the emission field.

These studies indicated the need for a more thorough examination of the shape and structure of the sonar field in the form of a detailed consideration of the particular function performed by the *pneumatized cristae maxillares* and other sound-insulating structures as part of the emission

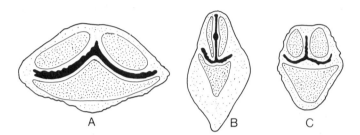

Fig. 93 Cross-section of the epiglottic spout to show differences in the shape of the epiglottic and cuneiform cartilages in *Hyperoodon ampullatus* (A), *Delphinus delphis* (B) and *Platanista indi* (C).

apparatus. This study also covers certain interesting bioacoustic features of the inspection behaviour*. The first step was to devise suitable investigation techniques, since *Platanista*, as a permanent swimmer, raises particular problems, and the literature suggests no appropriate methods for the study of sonar fields. On the basis of the available anatomical and acoustic material obtained from our own analyses and the relevant literature, an attempt was made to interpret the various emission fields of *Tursiops truncatus* and *Platanista indi*. As part of the receiving apparatus, the acoustic equipment already described was supplemented by a new Bruel and Kjaer 8103 hydrophone with a flat frequency response up to 140 kHz.

* Explanation of terms used to describe echolocation behaviour:
    1. *explore:* obtain a general acoustic image of the surroundings during normal swimming.
    2. *locate:* detect the position of an object by sonar.
    3. *inspect:* examine an already located object. In *Platanista indi* this gives rise to a special behaviour pattern.
The dolphin usually remains motionless with the object opposite its throat at an angle of approximately 30° to the axis of the rostrum. Head and neck are bent towards the object.

I. Signal Shape and Frequency Spectra

Details of the sonar signals of *Platanista* may be obtained from Pilleri *et al.* (1976a, b), but may be summarized as follows.

The position of the animal examined in relation to the hydrophone during recording is shown schematically in Figs 94 and 95. The signal sections were selected with a view to plotting polar diagrams of the dolphins' emission field. To obtain the two planes shown in the polar diagrams, the hydrophone was hung at the dolphins' swimming level in the tank. Generally speaking, the swimming level remained very constant, particularly in the cases frequently observed when the dolphins touched the bottom of the tank with their flippers while swimming. In this way, swimming along an even course on its side, an animal would move past the hydrophone with its back and its belly alternately towards the hydrophone. This produced the dorsal and ventral signal curves used to plot the polar diagrams. In this process, the hydrophone was always the fixed point and the animal the mobile point in the system. If the hydrophone was placed higher, we received signals from the lateral section of the emission area when the animal passed underneath (Fig. 96). In addition, special behavioural situations such as inspecting behaviour (Figs 97, 98) were investigated. Many more recordings than are described here were taken. The polar diagrams were arrived at from a combination of all recordings.

A. Findings

Recording No. 1: V 26_{12} male, ventral sector of the sonar field (Fig. 94).

Signal amplitude: In front of hydrophone position A, the amplitude amounts to 0·5 units in the oscillogram. At point A, the reading is 2·6 units, between A and B 6·3 at B 5·1 and at C 2·2 units. Halfway between C and D, the direct signal is no longer discernible.

Signal shape: The clicks all have $1\frac{1}{2}$ to $2\frac{1}{2}$ periods. From point A onwards, the signal is asymmetrical to the zero line, giving a ratio $+2$ to $-4·5$ (1 : 2·25). The third semi-period consistently shows a double peak throughout the entire click series.

Signal duration: Average signal duration 50–70 μs.

Spectrum: The frequency spectrum shows three spectrum peaks, which are fairly clearly common to the entire click series. Between A and B, the

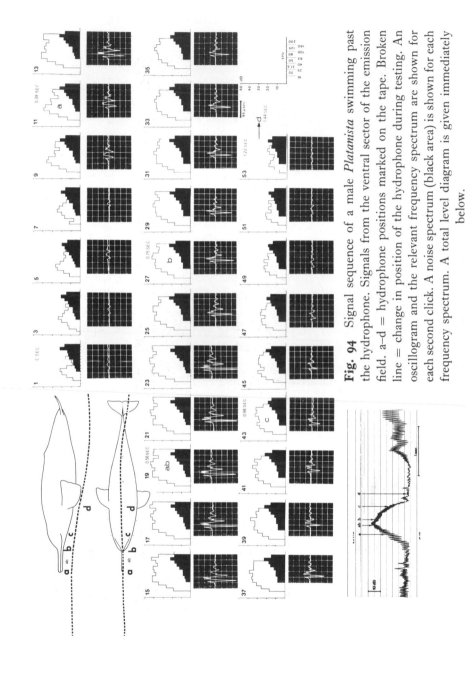

Fig. 94 Signal sequence of a male *Platanista* swimming past the hydrophone. Signals from the ventral sector of the emission field. a–d = hydrophone positions marked on the tape. Broken line = change in position of the hydrophone during testing. An oscillogram and the relevant frequency spectrum are shown for each second click. A noise spectrum (black area) is shown for each frequency spectrum. A total level diagram is given immediately below.

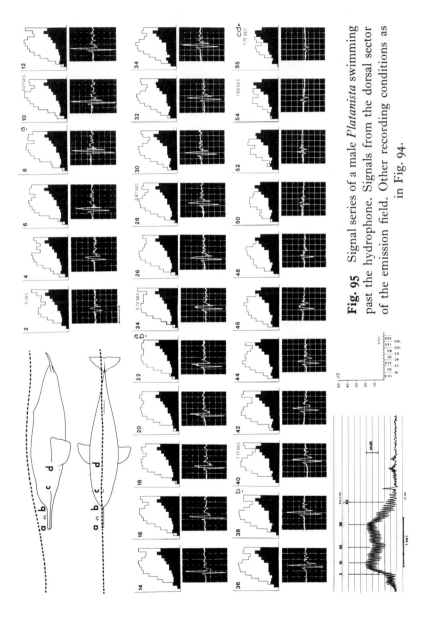

Fig. 95 Signal series of a male *Platanista* swimming past the hydrophone. Signals from the dorsal sector of the emission field. Other recording conditions as in Fig. 94.

peaks are at 25–30 kHz (1.M), 50–80 kHz (2.M), 160–200 kHz (3.M). In the caudal direction the highest frequencies are the first to decline between B and D.

Recording No. 4: V_{26} 4 male, dorsal sector of the sonar field (Fig. 95).

Signal amplitude: In front of position A, the amplitude amounts to 2·1 units in the oscillogram. At A, it is 7·8 units, and between A and B 5·3 units. At B it is 6·2 units and at C about 0·8 units. Sharp signal drop after B.

Signal shape: The clicks have 2–2½ periods. The asymmetry, at a maximum of 1 : 1·6 is lower than in recording No. 3, and is restricted to a small number of clicks. In position AB and after B, the third semi-period has two peaks.

Signal duration: The average signal duration is 50 μs.

Spectrum: There are two spectrum peaks, with a dominant at 50–63 kHz. There is a second peak at 160–200 kHz.

Recording No. 7: V 26_{15} female, lateral sector of the sonar field (Fig. 96).

Signal amplitude: Before A 3·7 units, at A 5·8 units, at AB 6·6 units. At B, the value is 6·6 units and 0·8 at C. The direct signal disappears between C and D.

Signal shape: Two periods are observed. The asymmetry between positive and negative semi-periods amounts to a maximum of 1 : 2·3. From just before B until just after B, the third semi-period is split.

Signal duration: The average signal duration is 40–50 μs.

Spectrum: There are two spectrum peaks, a dominant at 63–80 kHz, with a second peak at 200 kHz. A sharp drop occurs just before position C.

Recording No. 8: V 26_{28}. The male animal hovers on his side, touching the hydrophone cable with the end of his rostrum, which is 10 cm above the neoprene transducer head. Analysis of 10 clicks (Fig. 97).

Signal amplitude: Fairly constant throughout all clicks analysed (fluctuates by ±7% round the mean value).

Signal shape: All clicks show two periods. There is a maximum asymmetry of 1 : 1·4. All clicks have a single-peaked third semi-period.

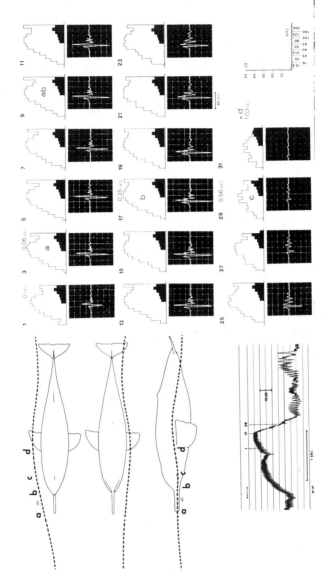

Fig. 96 Signals from the lateral sector of the emission field of a female dolphin. Recording conditions as in Fig. 94.

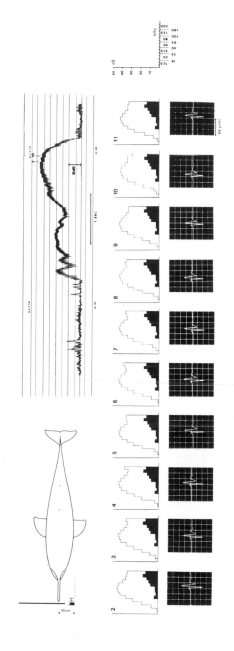

Fig. 97 Ten consecutive clicks emitted by the male *Platanista* in the lateral sector of the emission field. The animal remained motionless on its side, touching the cable 10 cm above the hydrophone with the point of its rostrum. Note the regularity of the oscillograms and frequency spectra.

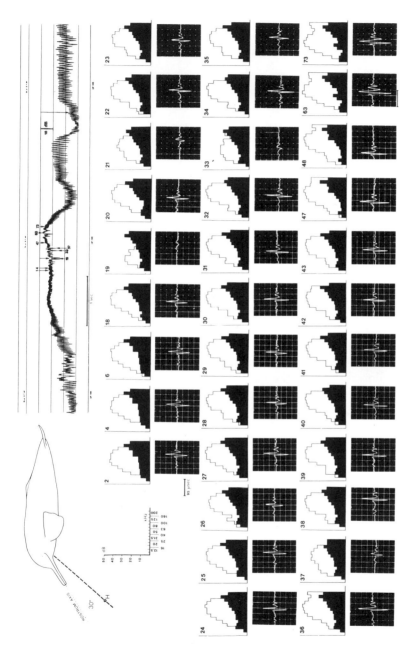

Fig. 98 Click series of the male *Platanista* in inspection behaviour. The dolphin remained motionless with the hydrophone opposite its throat at an angle of approximately 30° to the axis of the rostrum. Note the dominant frequency at 80 kHz.

Signal duration: The average signal duration is consistently 40 μs.

Spectrum: There is a clear spectrum peak at 63 kHz. The signal remains constant as long as the animal remains motionless at the hydrophone.

Recording No. 9: N 26$_{21}$ male during inspection behaviour, with head obviously bent, the hydrophone being anteroventral, 30° to axis of rostrum (Fig. 98).

II. Comparison of the Signals of the Male and Female Dolphin

A. Signal shape

The signal asymmetry distribution is different for the two animals. The male's ventral and lateral signals display the greater degree of asymmetry and symmetry respectively. The lateral asymmetry is not uniform in the two situations described. In one situation, the animal swims past the hydrophone; in the other case, the dolphin remains motionless near the hydrophone, touching the cable directly with the end of his rostrum (see Fig. 97). The female, on the other hand, has a highly asymmetrical lateral signal. The asymmetry of the ventral and dorsal signals is comparable with that of the dorsal signal of the male.

B. Spectrum

Figure 99 shows that the position of the dominant frequency ranges is very similar for both animals. It is clear from rough observation that the dominant peaks are somewhat lower for the male than the female in all ranges of the emission field. With the higher subdominants, the resolution in the analyses is not clear enough for any further differences to be perceived. In detailed examination, however, in the female the position of the dominant peak is much more sharply defined than for the male, whose spectrum peaks appear broader.

The ventral signal of the male in normal swimming is clearly differentiated from the female's signal in all field ranges. Three peaks can be observed in both ventral recordings analysed for the male. Moreover, it is not possible in both cases to indicate an unequivocally dominant frequency range. For the female, on the other hand, the dominant can be

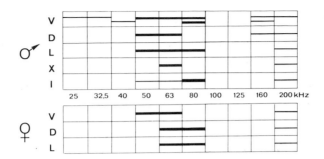

Fig. 99 Diagram of dominant frequencies (spectrum peaks) of the sonar signals of dolphins in different positions in relation to the hydrophone. The thick and thin lines show the dominant and sub dominant peaks respectively. V = ventral, D = dorsal, L = lateral, X = dolphin remaining in side position at the hydrophone, I = during inspecting behaviour.

more or less precisely pinpointed in the framework of the frequency uncertainty.

C. Discussion

A comparison was made of two recordings, both of which represent the throat signal of a normally swimming animal. For one recording, the hydrophone was hung in the middle of the tank, about 60 cm from the bottom (Fig. 100; clicks 8, 9); for the other recording, the hydrophone was placed at the swimming height of the dolphin, about 10 cm above the tank bottom (Fig. 101). The double peak in the third semi-period, already described, is shown only by clicks recorded with the hydrophone in the latter position. The phenomenon is therefore to be attributed to a superimposition of the echo from the tank bottom over the direct signal.

The clicks from all positions in the emission field are uniform in shape but not in spectrum. There are about two oscillations in the dominant frequency. Even in particular behaviour situations, such as inspection behaviour, only the dominant frequency changes slightly, but not the number of signal periods. The shift of the dominant frequency in inspection behaviour is presumably due to a change in the topographic position of the larynx caused by pronounced bending of the head when an object is being inspected (Pilleri *et al.*, 1976a, b).

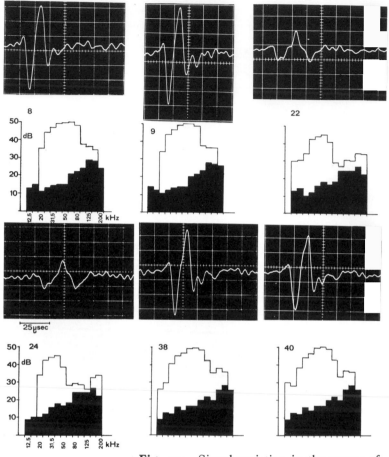

Fig. 100 Signal variation in the course of an exploratory head movement. The dolphin swam directly towards the hydrophone from a distance of about 3 m. Clicks 8 and 9 originate from the ventral sector, clicks 22 and 24 from the axial scotoma and clicks 38 and 40 from the dorsal sector of the emission field. The level curve shows the transition of ventral to dorsal field via the axial level discontinuity. The frequency spectra were corrected for the equipment frequency response and with a drop of 6 dB/octave to offset the change in the bandwidth of the 1/3 octave filters.

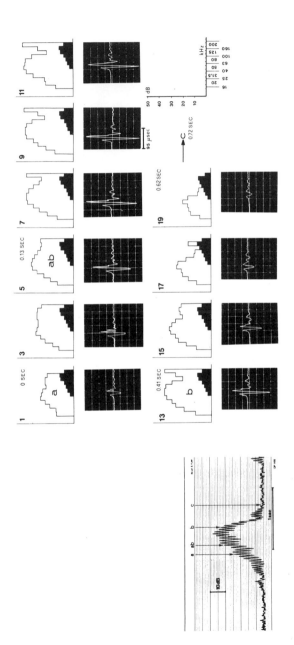

Fig. 101 Signal series of a female *Platanista* swimming past the hydrophone. Signals from the ventral sector of the emission field.

III. Directionality of the Emission Field

A. Methodology

As the dolphin swims past the fixed hydrophone, there is naturally a continual variation in distance between the sound source (larynx) and the hydrophone. If we consider the level of our recorded signals, it seems clear that the level variation observed is roughly speaking due solely to the variation in distance of the sound source from the hydrophone, since it occurs regularly and according to expectation. There is only a slight difference in the levels of consecutive clicks. It was possible to reconstruct the path followed by the hydrophone along the dolphin's body by means of marks made on the tape and by recorded comments. This made it possible to identify the individual clicks with the dolphin's position in relation to the transducer at the time the click was emitted. Figure 102 shows our subsequent procedure. Assuming that the sound emitted by the animal was propagated in a straight line and that it decreased in intensity as the square of the distance from the sound source, the following relation for the sound pressure level was obtained:

$$L_r = L_{r^\circ} - 20 \log \frac{r}{r^\circ}$$

where L_{r° represents the measured sound pressure level, r° the known radius from the sound source to the transducer and r the wanted radius.

Thus the sound pressure levels of the individual clicks measured at various distances from the sound source could be calculated with reference to the same distance $r = 1$ m from the larynx and subsequently compared.

The sound levels referred to equidistance lie on a circle with the larynx at the centre. The data thus obtained could now be plotted against the angles to the axis of the rostrum, providing the polar diagrams shown in Figs 103 and 105. The diagrams were calculated for eight different frequencies in each case, which covers the entire frequency range relevant to sonar.

If the emitter is taken to be an effective directional sound-reflecting surface of diameter 12 cm (maximum aperture of cristae maxillares), then for the higher frequencies contained in the sonar signal we must base our calculations on a linear rather than a quadratic decrease in intensity as a function of distance from the emitting surface. If the wavelengths of the high frequencies occurring in the signal become smaller than the size of

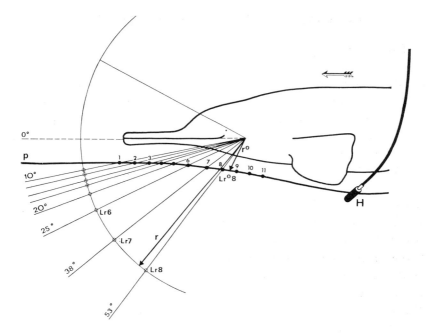

Fig. 102 Schematic representation for calculation of the standard sound pressure level ($r = 1$ m) from the measured values. The arrow shows the direction in which the dolphin swam past the hydrophone. The positions in which the hydrophone was when clicks Nos 1–11 were received lie on p. The sound levels L_{r° (1–11) measured were converted to standard sound pressure levels L_r (1–11) with $r = 1$ m. H = hydrophone.

the emitter, the latter can no longer be regarded as a point source and the sound propagation law given above must be converted to:

$$L_r = L_{r^\circ} - 10 \log \frac{r}{r^\circ}$$

which is equivalent to a linear decrease in intensity with distance from the sound source.

To verify this approach, one of the polar diagrams was calculated using the second formula. The diagram obtained, taking account of the errors which may have crept in as a result of inaccurate coordination of the distance and time data obtained from the recording, showed so little difference from the result calculated on the basis of $1/r^2$, that it need not be reproduced at this point.

B. Results

Figure 103 A–D shows the ventral and dorsal sectors of the emission field of the male dolphin in the median plane. All the frequencies of the ventral and dorsal polar diagrams shown in the figures are highly directional.

The signal level is highest at an angle of 15–25° to the rostrum axis, both dorsally and ventrally. At lower angles, the 80 kHz signal decreases at a rate of 2·5–5 dB per degree. Greater angles to the rostrum axis show level drops which are smaller than 1 dB per degree at 80 kHz. In the ventral field, the signal at 80 kHz has decreased by 25 dB at an angle of 60° to the rostrum axis. In the dorsal field, the corresponding decrease is about 30 dB. The level drops of the other frequencies shown are within similar limits.

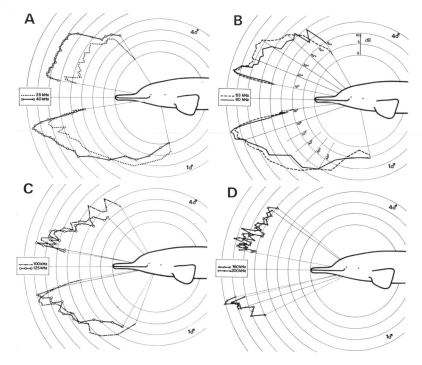

Fig. 103 (A–D)   Polar diagram of the sonar field of the male dolphin in the median plane. Since *Platanista* is a side swimmer, this plane is parallel to the tank floor. The measurement values used to plot the diagram were obtained with one-third octave filters of the indicated mean frequencies.

A different recording technique had to be used to investigate the central field range around the peak axis. In normal swimming, *Platanista* makes very frequent exploratory movements with the head in the median plane. A hydrophone was hung at the swimming level of the dolphin to record the sonar signals of the animal as it swam directly towards the hydrophone. Two time marks were made on the tape indicating complete bending and stretching of the head. In between, clicks from the axial sector of the emission field around the rostrum were recorded.

Figure 104 A–D shows the corresponding data obtained for the female. A central "scotoma" of the emission field in the beak region is obtained by combining Figs 102 and 103 which are mutually complementary.

Finally, if we look at the lateral diagrams (Fig. 105 A–D), we see a similar directional characteristic for both the male and the female animal. The signal level is highest at all frequencies at a lateral angle of 15–30° from the axis of the rostrum. In the lateral field near the rostrum, the

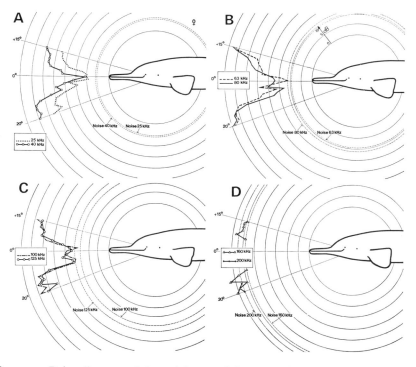

Fig. 104   Polar diagram of the axial part of the sonar field in the median plane. Female Indus dolphin. This diagram supplements Fig. 103. The broken circles represent the apparatus noise levels.

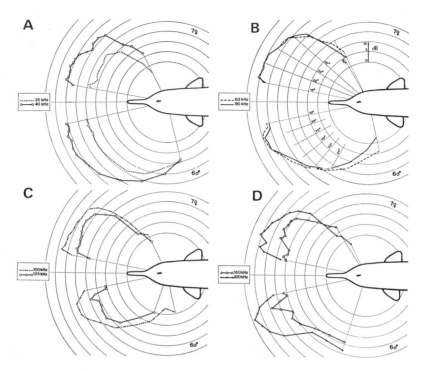

Fig. 105 Polar diagram of the lateral sector of the sonar fields of the male and the female Indus dolphin. The plane indicated is vertical to the water surface for the swimming animal.

signal level for the female between 10 and 50° shows a relatively slight drop (10 dB) at 80 kHz, the corresponding figure being only slightly higher for the male (13 dB). For the female the level drop rises to about 1 dB per degree in the rest of the field sector. Other recordings of lateral patterns suggest that the right and left lateral fields of an animal are symmetrical. Hence the examples given in Fig. 105 are to be regarded not as representative of the left and right side of the body but as characteristic of a different animal in each case.

A combination of the lateral and median fields of one animal displays an emitting space with high directivity in the forward direction, the emitting energy being roughly the same in the ventral, dorsal and lateral sectors. A clear discontinuity in the sound level is observed only around the axis of the rostrum. In all parts of the emitting space, there is a more or less pronounced drop in signal level at angles of greater than about 30°

to the rostrum axis. The drop is not uniformly sharp, as can be seen clearly from the polar diagrams. On the basis of the available data, it is not possible to determine whether true individual differences exist or not.

At high frequencies, the polar diagrams are extremely uneven, particularly in the dorsal but also in the ventral area of the emitting space. Radiation is more uniform in the entire space at low frequencies.

IV. The Function of the Cristae Maxillares in
Platanista indi

A. Methodology

The previous research (Pilleri *et al.*, 1976a, b) indicates that the larynx is the only source of the sonar sounds emitted by *Platanista indi*. During sound production, the larynx is inserted into the opening of the palato-pharyngeal sphincter and is gripped firmly by this muscle.

The fact that a discontinuity in the sound level was observed in the central space of the dorso-ventrally extended sound field in the area around the rostrum axis, and the finding that the pneumatized cristae maxillares constitute an excellent sound screen prompted the authors to adopt the following hypothesis.

The dorsal and ventral sectors of the emitting space of *Platanista indi* do not both receive sounds directly from the larynx. The dorsal part of the sound output pupil is located in front of the aperture of the larynx, which projects forward dorsally into the palatopharyngeal sphincter, and thus receives the direct signal. In the ventral part of the emitting space, exclusively larynx signals occur, which are reflected on the air buffer of the cristae maxillares (see Pilleri *et al.*, *loc. cit.*).

Since the dolphin's signal was invariably recorded in both the ventral and the dorsal part of the emitting field, the signals must be assumed to originate from the same source. If this hypothesis is correct, the signal recorded in the ventral part of the sonar field should be delayed, as owing to reflection from the cristae maxillares it has further to travel than the signal observed in the dorsal part of the field.

In order to verify this, the following experimental arrangement was set up. Holes were drilled at regular intervals of 15 cm in a wooden stick 2 m long. This was fixed above the surface of the water radially to the dolphin's normal swimming circle. Two hydrophones were then hung on the stick, one at each end. The distance between the two hydrophones was set at 165 cm, so that one hydrophone was in the dorsal and the other in the ventral sector of the emission field when the dolphin swam past. As soon as the dolphin had assumed his normal swimming pattern, the test rig was so arranged that the animal headed directly for the mid-point between the two hydrophones on each circuit. In this way, at the time of recording, the rostrum-skull axis was vertical to a straight line between the two hydrophones, which were equidistant from the dolphin's head. The best situations were marked by a centrally positioned observer on a third sound track. The two first tracks of the tape were supplied with the dolphin signal using two identical recording chains. The equipment used was a Nagra-IV-SJ triple-track magnetic recorder with a bandwidth of 35 kHz at 38 cm/s. The tape speed was reduced to one-eighth on playback. The time difference between the two channels was measured on a dual beam storage oscilloscope.

B. Results

Time delay of signal in the ventral emission field:
Figure 106 provides a schematic description of the test procedure. In the starting position, both hydrophones received signals from the ventral area of the emission field. The signals show virtually no time lag (oscillogram, Fig. 106: *1*). If the animal approached the two hydrophones (Fig. 106: *2*), one hydrophone (H 2) received signals from the ventral sector and the other (H 1) received signals from the dorsal sector of the emission field. The relevant oscillogram shows a time lag of 150 μs for both channels. If the animal swam past the hydrophone assembly, hydrophone H 2 came into the dorsal sector of the field, whereas hydrophone H 1 was already outside the directional field and received only echoes.

In stage 2 of the swimming pattern, past the hydrophones, the signal from the ventral sector of the emission field therefore arrived later than the signal from the dorsal sector.

V. Discussion Concerning Sections III and IV

The information contained in the relevant literature concerning the directionality characteristics of the sonar fields of dolphins is scanty and

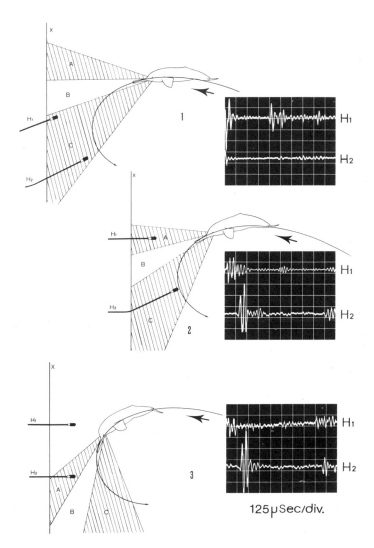

Fig. 106 Test assembly to determine the time lag of signals from the dorsal (A) and ventral (C) sectors of the emission field. B = axial sector of the emission field. Overhead view of the tank and the dolphin swimming on its side towards the tank wall (X). Both hydrophones (H1 and H2) are hanging vertically 165 cm apart. On the right-hand side, the oscillogram from situation 1 and two oscillograms from situation 2. Signal from the tape, filtered at 30 kHz (1/3 octave).

largely incomplete. Norris *et al.* (1966) described a horizontal plane through the emission field of *Steno bredanensis*. This work contained no data on the dorso-ventral plane, in which we would have been particularly interested. Pilleri *et al.* (1978) have already pointed out that sonagrams are not suitable for the investigation of sonar signals. Since the diagrams given by Norris were based on sonagrams and sound pressure r.m.s. values, they are scarcely comparable with our polar diagrams. A study by Evans (1973) concerning *Tursiops truncatus*, on the other hand, offers a better basis for comparison. Although Evans' polar diagrams are not uniform and cover only a narrow level range, there are very clear differences between the emission fields of *Tursiops truncatus* and *Platanista indi*. For *Tursiops*, the signal in the horizontal plane (lateral field) was strongest in a straight line in front of the rostrum. *Platanista*, on the other hand, showed a decline in sound level of more than 20 dB below the maximum sound level at this point in the lateral field. The most intensive signal occurred in an angle of 15–20° right and left of the rostrum. In the vertical plane, the differences in the shape of the emission fields of both species are even more apparent. Norris *et al.* (1976) found that a *Tursiops* whose eyes had been covered with non-transparent suction cups could only detect pieces of fish located above his rostrum line in the water. The food floating below the rostrum line was ignored. The authors concluded that *Tursiops* has only a dorsal emission field.

On the basis of recent investigations on the sonar emission field of *Tursiops truncatus* (Pilleri *et al.*, in preparation) it was proved that Norris' conclusion is erroneous.

Examination of the emission field in the vertical (=dorso-ventral) plane shows clearly that, in addition to the dorsal lobe, *Platanista* also shows a marked ventral lobe with a sound level drop of up to 25 dB around the rostral axis (Figs 103 A–D, 104 A–D).

Can these differences in the shape of the emission fields of both species be attributed to differences in the anatomical substrates? Taking a sagittal cut through the head of both types, we find two fundamental anatomical differences in the region of the sound source (larynx). In *Tursiops truncatus*, the powerfully developed pterygoids with their hamuli extend a long way in the caudal direction until about the level of the front edge of the basi-occipital bone. In the process of sound emission, the epiglottic spout is thrust into the palatopharyngeal sphincter, and constricted by it to form a solid unit. The paired pterygoid sinuses form a latero-ventral screen for the acoustic signals emitted by the larynx.

In *Platanista indi*, the pterygoid sinuses are much smaller, and the cartilaginous hamuli extend in the caudal-basal direction less far than in *Tursiops truncatus*. Anatomically and topographically speaking, they are scarcely capable of constituting a lateral screen against the larynx signals. This function is fulfilled by the paired nasopharyngeal pouches. They follow the pterygoid sinuses directly and surround the nasopharyngeal sphincter on both sides. The topography and structure of these pouches and their relationship to the Eustachian tube has already been described in detail (Purves and Pilleri, 1973; Pilleri *et al.*, 1976). This provides *Platanista* also with a complete sleevelike screen of the emitting apparatus (larynx with palatopharyngeal sphincter), which only leaves the rostral aperture of the larynx open during phonation. The nasolaryngeal pouches are relatively thin-walled and extensible in the rostral section around the sphincter. It is also the frontal section, between the sphincter and the stylohyal, which is displaced dorsally and compressed during swallowing, accompanied by lowering of the larynx and maximum extension of the pharynx. We may assume that the air escapes from this part of the naso-laryngeal chambers during swallowing. In this situation phonation is physiologically impossible (see Pilleri *et al.*, 1976).

The remaining part of the guttural pouches is located ventro-mesially in the larynx and the caudal pouch pole extends almost to the first segments of the trachea. Attention has already been drawn to a possible dorsal screening by the oesophagus. We cannot tell from Evans' measurement diagrams (1973) how far the sonar field extends dorsally in *Tursiops truncatus*. In *Platanista*, the dorsal field appears to be displaced dorsally about 10–15° from the rostrum. The different anatomical characteristics of the pterygoid sinuses of both species might explain this displacement. We see from examining the skull that the dorsal edge of the pterygoid sinuses in *Tursiops* coincides precisely with the horizontal base of the vomer. In *Platanista*, this edge forms a ramp rising towards the rostrum which may cause the 15° slope in the dorsal emission field.

The findings concerning the time difference between the dorsal and ventral signals suggest that the *sonar signal in the ventral lobe of the emission field is a larynx signal reflected by the pneumatized maxillary crests*. In fact, therefore, this would give two sound radiating regions in *Platanista*, each of which forms its own sound field: (a) the region of the larynx and the palatopharyngeal sphincter, which forms a functional unit and emits rostro-dorsally via the vomer, and (b) the pneumatized cristae maxillares, which reflect the same signals rostro-ventrally (Fig. 107).

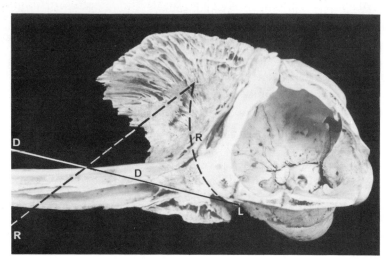

Fig. 107   A theoretical diagram of the direct signal (D) and the signal reflected (R) from the larynx (L) on a macerated skull of *Platanista indi*.

In the zone where the two fields intersect appears the sound pressure discontinuity, which causes an axially located acoustic "scotoma". As explained above, *Tursiops* has only a single, rostral emission field owing to the massive ventral screening of the sound apparatus by the paired pterygoid sinuses.

Platanista constantly nods its head during swimming. These exploratory head movements in the median plane of the body might be related to the central "scotoma" in the emission field. *Platanista* might use this very pronounced head movement in order to eliminate this "blind spot".

Previous studies (Pilleri *et al.*, 1978) indicated that *Platanista* locates an object at an angle of 25–30° ventrally to the rostrum axis and not in the immediate direction of the rostrum. In this process, the head is bent at such an angle that transverse creases appear in the throat region. Bearing in mind the osteological configuration of the *Platanista* skull, it is clear that in this position the opening of the pneumatized cristae maxillares is focused directly on the object under examination. Both the central axis of the sound funnel formed by the two cristae and the free edges of their dorsal combs are in the rostro-ventral direction, at an angle of 30° to the rostrum axis (Fig. 108).

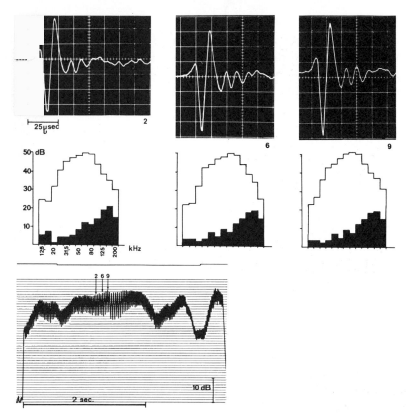

Fig. 108 Oscillograms and frequency spectra of female Indus dolphin clicks recorded in inspection behaviour. The clicks are plotted in the level diagram. During recording, the hydrophone was hung in the middle of the tank, 50 cm from the nearest reflecting surface. This prevented overlapping of direct signal and echo. The frequency spectra were corrected for the equipment frequency response and with a drop of 6 dB/octave to offset the change in the bandwidth of the 1/3 octave filters.

These functional and anatomical factors encourage the assumption that the *inspection signal can only emanate from the crista*. The vital significance of the crista signal for the Indus dolphin can be demonstrated by its feeding behaviour. When *Platanista* locates a living fish from a distance of several metres, it heads for the fish with its head bent, so that the throat region is constantly turned towards the fish. At a distance of about 50–70 cm from the prey, the dolphin opens its beak wide by abduction of the lower jaw. As soon as the fish has passed between the gaping jaws,

the dolphin snaps them shut by rapid adduction of the lower jaw and passes
the fish along the groove of the beak to the oral cavity.

It was observed that the dolphin still emits sonar sounds when its beak
is wide open. The signals do not cease until the dolphin swallows, i.e.
when the fish slips into the oesophagus over the lowered larynx (Pilleri
et al., 1976).

If we follow the eating process, it is clear that after *Platanista* has
located the fish, it is operating only with the ventral part of the emission
field, i.e. using the crista signal (Fig. 109). Owing to the axial scotoma, it
would be physiologically pointless for *Platanista* to copy the marine
dolphin by taking a bearing on its prey in the rostrum axis. The dorsal

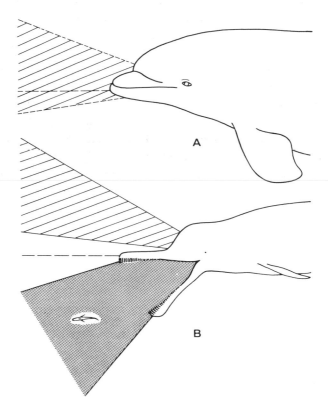

Fig. 109 (A–B) Schematic representation of the emission fields of *Tursiops
truncatus* (A) and *Platanista indi* (B) in the median plane. When hunting,
Platanista indi directs the ventral sector of the emission field at the fish, which is
therefore located by the crista signal.

emission field is also useless for hunting, since it extends in front of the melon, quite outside the mouth region.

A last question is raised with regard to the phyletic development of the cristae maxillares and the closely connected ventral part of the sonar field of *Platanista*.

It is well known that the blind Indus dolphin is a side swimmer. It was assumed that, in the history of the species, this particular swimming style would constitute a secondary adaptation caused by the degeneration of the eyes (Pilleri, 1974, 1975). The cristae maxillares, which are still lacking in a foetus of 45 cm body length, were also a late differentiation (Pilleri and Gihr, 1976). The anatomical study of the head of the adult *Platanista indi* has, *inter alia*, revealed the absence of the vestibular air-sacs which are present in marine species and other river dolphins.

Moreover, pneumatized cristae maxillares have been described in fossil Platanistoidea only for *Ischyrorhynchus vanbenedeni* (Pilleri and Gihr, 1979).

The cristae of *Zarhachis flagellator* are very low, not pneumatized and form an angle of 90° to the plane of the maxilla. *Zarhachis* is not to be classed as a precursor of *Platanista* but as a parallel species from the upper Miocene. Therefore the ventral part of the sonar field should likewise be a secondary functional differentiation which developed *pari passu* with the evolution of the cristae. According to Pilleri's hypothesis, the loss of the eyes would have modified the original swimming mode, caused the animal to adopt the side position, as well as activating the groping function of the pectoral fin.

VIII.   The Ear of Cetaceans

There are no conspicuous external ears in whales and dolphins but the cartilages and lumen are present below the blubber layer. Cetologists in the United States, using electrophysiological techniques, have concluded that the external ears and considerable parts of the middle ears are degenerate, despite the fact that the latter are almost identical with those of the Squaladontidae or shark-toothed cetaceans which lived during the Oligocene and Miocene period more than 20 million years ago. There is evidence from endocranial casts of the brain and the presence of large olfactory foramina in the fossil skulls that these ancient cetaceans possessed a sense of smell. There is no trace of olfactory bulbs nor nerves in modern cetaceans except in the large plankton-feeding baleen whales. It seems odd therefore that this group of animals should have lost all trace of the structures relating to the sense of smell which is inadequate for a mammal under water, whilst retaining structures relating to hearing which are equally useless! Norris (1968) has postulated that sound is received through the lower jaw and judging by the number of times that his diagram has been reproduced internationally the idea has been given world-wide credence.

However the Russian acoustician Romanenko (1976) stated:

> "The placement of two wideband noise radiators with a level of about 120 dB at different points on the head of the dolphin (the radiators are LZT ceramic spheres 30 mm in diameter) has verified that the sharpest response, both locomotive and acoustic, of the dolphin to noise generation is observed when the radiators are placed in the vicinity of the outer auditory passages. The attachment of radiators to the frontal adipose cushion and the lower jaw elicits practically no response. This result suggests that sound is transmitted to the inner ear of the dolphin mainly through the outer auditory passages".

In his "Textbook of Sound" Wood (1955) states "It is important to note that electrophysiological experiments on nerve impulses made on

anaesthetised animals do not necessarily represent in detail the normal
conscious state of the animal's mind", and Wever (1949) refers to experiments on nerve response and refractory periods in other parts of the body,
e.g. nerves associated with leg-muscles, and nerves in teeth, responding
faithfully to mechanical vibrations of audiofrequency.

In the experiments of McCormick *et al.* (1970) in the bottle-nosed
dolphin, *Tursiops truncatus*, using 20 kHz sound stimulation, all the areas
of the head found to be sensitive at the level of the auditory colliculus are
also innervated by massive branches of the trigeminal nerve, especially
the teeth. It follows from Wever's remarks that potentials measured at the
trigeminal nucleus would be equal to, if not greater than, those measured
at the auditory colliculus. It would seem important in view of Romanenko's
results to compare the nature of the auditory collicular potentials with
those from the trigeminal nucleus to find out how much sound processing
is carried out at the level of the cochlea. It would seem that so far this has
not been done and it illustrates the importance of a knowledge of cetacean
anatomy in electrophysiological experiments.

I. External Auditory Meatus and Auricular Muscles

A. *Tursiops truncatus*

The external auditory meatus of *Tursiops* is recognizable as a minute
orifice on the side of the head, 5·5 cm behind and 1·5 cm ventral to the
hinder corner of the eye (Fig. 110: EAM). The meatus penetrates the
blubber in a horizontal and slightly antero-posterior plane as a pigmented
tube surrounded by fibrous tissue, and contains a lumen just large enough
to admit a fine bristle. After passing through 2·35 cm of blubber, the meatus
becomes encased in a conical mass of fibrous tissue associated with a number of auricular muscles and the distal end of the auricular cartilage
(MOA, MZA).

Immediately anterior to the sternomastoid and mastohumeralis muscles
(Fig. 110: MSM, MMH), a reniform fibroelastic lobe (AH) projects
from the ventral aspect of the mastoid process of the squamosal. This lobe
has an anteriorly directed flexure and supports, along its ventral margin,
a band of cartilage of a characteristic shape, consisting of a sigmoid, semitubular strip 1–2 mm thick and 6–7 mm wide, the concavity of which
contains the lumen of the external auditory meatus. Medial to the conical

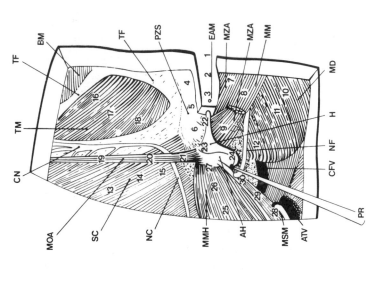

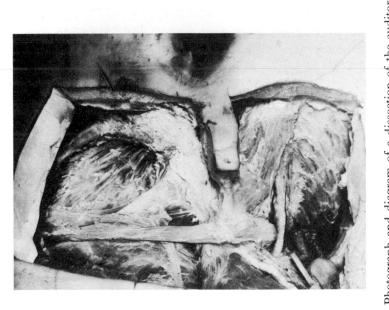

Fig. 110 Photograph and diagram of a dissection of the auditory region of *Tursiops truncatus* to show the relationship of the external auditory meatus to the eye and the underlying auricular muscles. The figures in the diagram represent points at which sound conduction were measured (see Table 10). AH = antihelix, ATV = anterior temporal vein, BM = blowhole muscle, CFV = common facial vein, CN = second cervical nerve, EAM = external auditory meatus, H = helix, MD = digastric muscle, MM = masseter muscle, MMH = mastohumeralis muscle, MOA = occipitoauricularis muscle, MSM = sternomastoid muscle, MZA = zygomaticoauricularis muscle, NC = occipital branch of cervical nerve, NF = facial nerve, PR = probe, PZS = zygomatic process of squamosal, SC = splenius capitis muscle, TF = temporal fascia, TM = temporal muscle.

mass of fibrous tissue referred to above, the lumen of the meatus passes ventrally in a trough enclosed by the cartilage. The cartilage and the meatus wind anteriorly and ventrally for 3·7 cm, then posteriorly and dorsally and mesially towards the tympanic bulla (Fig. 111: TB). This inner part of the lumen of the meatus varies from 2 to 3 mm in diameter, and is lined by brown pigmented epithelium.

From their interrelationships, it seems likely that the meatal lumen, the fibroelastic lobe, and the cartilage may be regarded as a single unit, and as the homologue of the pinna of terrestrial mammals. The cartilaginous trough containing the meatal lumen may be regarded as the helix (H) and the fibroelastic lobe as the antihelix (AH). The distal extremity of the helix has a small, mesially directed flange, which may be referred to as the tail of the helix (TH). At the proximal end of the helix, 9 mm from the bulla, a second cartilaginous flange forms a continuous band, completely enclosing the external meatus for 13 mm; this band may be referred to as the cartilage of the meatus (CM) (see Gray, 1964). A fleshy ligament covering the proximal 10 mm of the meatal tube connects the cartilage of the meatus with the processus conicus, posterior to the tympanic bulla.

The lateral extension and redisposition of the auricular cartilage from the typical mammalian form is associated with the lengthening of the auditory canal in cetaceans, and with the lateral extension of the mastoid process of the squamosal. Its subdermal position is associated with the proliferation of the panniculus adiposus, or blubber.

The helix, on its anterior aspect, is in relation to the ventral margin of the glenoid fossa with which the lower jaw articulates. On its posterior aspect it is in relation to the insertion of the sternomastoid and masto-humeralis muscles. The cartilage of the meatus passes anteriorly to the styloid cartilage (STC) and posteriorly to the posterior wing of the middle sinus and its fibrovenous plexus (FVP). The mandibular branch of the facial nerve (NF) emerges from the styloid foramen posteriorly to the proximal limit of the helix. On the antero-dorsal aspect of the bulla, a thick band of muscle is inserted into the processus tubaris of the periotic and the lateral margin of the ostium tympanicum tubae and Eustachian tube (ET). This tensor palati muscle (MTP) encloses, on its mesial aspect, the pterygoid air-sac and the fibrovenous plexus. The tympano-periotic bones, as is usual in cetaceans, are separated from the cranium by the peribullary system of air spaces (PBS).

Deep to the blubber and superficial fascia, a strap-like muscle originates in a strong tendon from the supraoccipital crest, and, passing ventrally

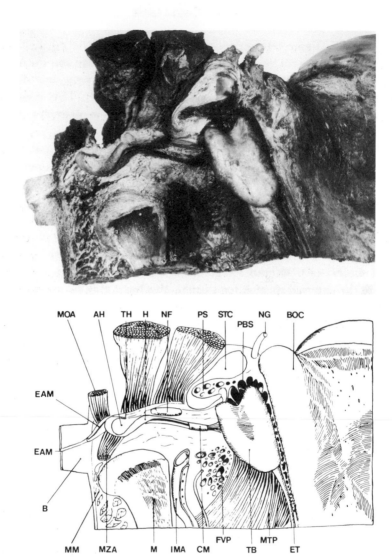

Fig. 111 Dissection and diagram showing the tympanic bulla and cartilaginous external auditory meatus of the bottle nosed dolphin, *Tursiops truncatus*. AH = antihelix, B = blubber, BOC = basioccipital bone, CM = cartilage of meatus, EAM = external auditory meatus, ET = Eustachian tube, FVP = fibrovenous plexus, H = helix, IMA = internal maxillary artery, M = mandible, MOA = occipitoauricularis muscle, MTP = tensor palati muscle, MZA = zygomaticoauricularis muscle, NF = facial nerve, NG = glossopharyngeal nerve, PBS = peribullary sinus, PS = posterior sinus, STC = styloid cartilage, TB = tympanic bulla, TH = "tail" of helix.

superficial to the splenius capitis, becomes gradually broader and more fleshy, and is inserted into the superior, distal extremity and the anterior face of the antihelix. This muscle is the "retrahens" of Murie (1873), the "occipitoauricularis" of Boenninghaus (1903), and the "auricularis externus" of Beauregard (1894) (Fig. 110: MOA). It is much stouter than the muscles described by these authors, and is richly innervated by small twigs from the lesser occipital branch of the second cervical nerve (CN). In *Phocoena*, the fleshy part of the muscle is relatively much longer than as depicted by Boenninghaus (1903).

Originating in an aponeurosis from the superficial fascia of the masseter muscle (MM), near the origin of that muscle on the zygomatic process of the squamosal, a fan-shaped group of small muscles passes postero-dorsally and is inserted fleshily into the anterior face of the antihelix. These muscles are the combined attolens and attrahens of Murie (1873), the zygomaticoauricularis of Boenninghaus (1903), and the auriculolabialis of Hanke (1914) (MZA). They are innervated by the facial nerve.

The general disposition of the meatus and cartilages on the two sides is similar, but it is noteworthy that in one of the specimens of *Tursiops* examined the fibroelastic lobe, or antihelix of the left side, was semi-discoid in appearance, and the sigmoid cartilage of the helix had a more deeply ventrally directed flexure than that on the right. In a dissection of the meatus of the pilot whale, *Globicephala melaena*, figured by Fraser and Purves (1954) there appears to be no ventrally directed flexure of the cartilage of the helix, but, instead, a pronounced dorsally directed flexure at its distal extremity. In a specimen of *Phocoena phocoena* described and figured by Boenninghaus (1903), the auricular cartilage has a dorsally directed flexure at its distal extremity. In Boas' (1912) figure of *Phocoena* there is no noticeable flexure of the cartilage at all. It is conceivable that contraction of the superior auricular muscle may substantially alter the shape of the helix, and that the differences in the flexure of the cartilage described by various authors may be attributed to the state of tension of this muscle at the time of dissection. The possible significance of voluntary control of the internal pinna will be discussed later, but it is of some importance to note at this stage that Hunter (1787), in describing the meatus of the porpoise, stated:

> "It passes in a serpentine course, at first horizontally, then downwards, and afterwards horizontally again, to the membrana tympani, where it terminates. In its whole length it is composed of different cartilages, which are irregular and united together by cellular membrane, so as to admit of motion, and probably of lengthening and shortening, as the animal is more or less fat".

The external auditory meatus was the first structure to be eliminated by some cetologists as having a role in cetacean hearing. However, Batteau (1968) demonstrated that the shape of the external ear in Man was important in the localization of sound especially in the higher frequency range above 6 kHz. He found that he made an error of 90° in the blindfold localization of the sound of jingled keys when the tips of his ears were bent down. We found that stretching the cartilage of the meatus in *Tursiops* could produce a phase-shift of 90° in a 100 kHz monotone without any apparent attenuation. When one considers the obvious functional significance of the auricular muscles in *Tursiops* one must conclude that the external auditory meatus is also functional. The semitubular nature of the meatal cartilage is such that when stretched, no stresses large enough to fracture the cartilage occur.

To quote Batteau: "in order to know, one must be able to construct the inverse to the transformation of perception". The proof is simple—if the inverse cannot be constructed then the correspondence between received information and that which gave rise to it cannot be made. He goes on to use this in explaining the "cocktail party effect":

> "By applying a second appropriate transformation amounting to a matched filter, the sound from a particular locale may be attended to, or selected by an increase of signal-to-noise ratio over other locales to which the transformation in use does not apply".

We consider that the external meatus of *Tursiops* is eminently capable of carrying out such transformations. Whilst admitting that the tissues surrounding the meatus can also conduct sound we found that the cartilage of the meatus is by far the best conductor.

B. *Balaenoptera physalus*

In this species the external auditory meatus (Fig. 112) may be observed as a lenticular opening 7 mm in diameter. Within the layer of blubber, which is approximately 10 cm thick in this region, it widens to about 1 cm and is lined by black, pigmented epithelium (EAM). At the termination of the blubber it narrows to about 5 cm, and finally becomes obliterated in a mass of fibrocartilage, the auricular cartilage (AC), whose exact shape has not been ascertained. The corium of the meatus emerges mesially from this cartilage, to which auricular muscles are attached, as a tightly stretched cord 6 mm thick, consisting of a mass of white fibrous and yellow elastic

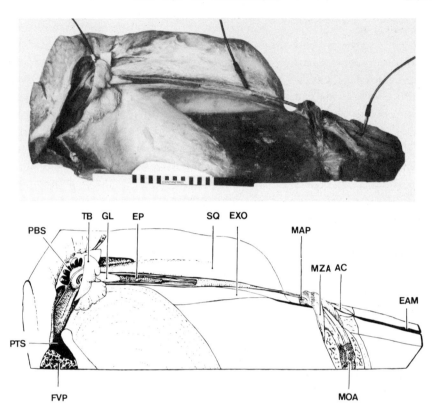

Fig. 112 Photograph and diagram of a dissection of the external auditory meatus and middle ear of *Balaenoptera physalus* the measal half of the tympanic bulla has been removed. The acoustic probes were used to measure the attenuation of sound waves in the meatus.

AC = auricular cartilage, EAM = external auditory meatus, EP = ear plug, EXO = exoccipital bone, FVP = fibrovenous plexus, GL = "glove finger", MAP = auricularis posterior muscle, MOA = occipitoauricularis muscle, MZA = zygomaticoauricularis muscle, PBS = peribullary, PTS = pterygoid sinus, SQ = squamosal bone, TB = tympanic bulla.

tissue arranged in longitudinal fibres running parallel with the long axis of the cord. This cord is 10 cm long and curves gradually ventral, but there is no sharp bend as depicted by Lillie (1910), Hanke (1914), and Reysenbach de Haan (1957). The cord is highly elastic, and retracts into the surrounding connective tissue when severed, so that it is necessary to carry out the bloc dissection in order to study its proper course. At the mesial end of the cord, the corium gradually widens to a maximum of 4 cm in diameter, and contains on its inner aspect the stratum germinativum, the stratum corneum, and the lumen of the proximal part of the external auditory meatus. This proximal part, 35 cm long, is filled by a large number of layers of keratinized epithelium forming the "earplug" (EP; Purves, 1955). At its mesial extremity the meatus is attached to the anulus of the bulla and, with the earplug, envelopes the pars flaccida or "glove finger" (GL) of the tympanic membrane (Lillie, 1910).

The whole length of the corium and meatus, from the mesial extremity of the cord to the tympanic anulus, lies in a deep groove between the exoccipital (EXO) and the squamosal bones (SQ), and is surrounded by loose connective tissue and oil sinuses. Enveloping the whole of this area, and adhering to the bones, is a great mass of dense, white fibrous tissue about 30 cm in thickness with tough, unyielding fibres forming a close reticulum throughout the mass. The corium and meatus are thus enclosed in a tunnel formed dorsally by bone and ventrally by the mass of white fibrous tissue.

The auricular muscles are very conspicuous, and vary from 3 to 8 cm in diameter. In addition to the muscles found in odontocetes, the occipito-auricularis and the zygomaticoauricularis, there is a small muscle that originates on the posterior, mastoid aspect of the squamosal, and by its insertion into the auricular cartilage could be an auricularis posterior (MAP). Another small muscle, originating on the superficial fascia of the occipitoauricularis, could be an auricularis anterior (MAA). The whole muscle complex is associated with a dense retial mass of blood vessels and fatty tissue, which probably indicates that the muscles are functional.

II. Middle Ear

A. Tympanic bulla

The external auditory meatus terminates mesially at the lateral opening, the tympanic anulus, of an ovoid capsule of bone, the tympanic bulla, which is situated on the ventro-lateral aspect of the posterior extremity of the cranium. The tympanic bulla whose general shape may be seen in Fig. 111 (TB), is sufficiently well known not to require detailed description. It may be said to consist of a mesial wall and a lateral wall which are joined ventrally and posteriorly, but are separated dorsally and anteriorly by the tympanic cavity. The mesial wall is characterized by its stout petrous nature (the bone in rorquals being up to 3 to 4 cm thick) and by the smoothness of most of its surface. Both of these features are associated with its contiguity with air sinuses. The lateral part is much thinner and has a roughened outer surface; these features are associated with its contiguity with the surrounding fibrous layer, which is about 12 cm thick in a large rorqual and is strongly adherent to the lateral surface of the bone. The bulla in whalebone whales, half of which has been cut away (Fig. 113)

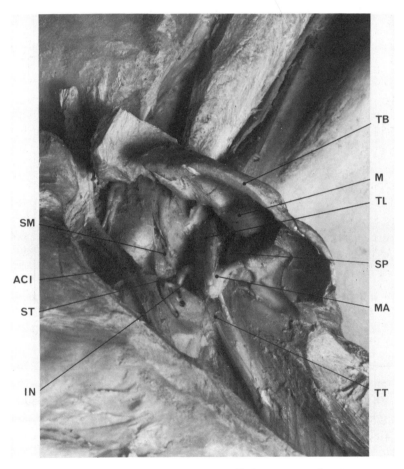

Fig. 113 Photograph of the middle ear of *Balaenoptera physalus* the ventral wall of the tympanic bulla has been removed and the cochlea partially dissected to show the spiral labyrinth.

ACI = internal carotid artery, IN = incus, M = enigmatic muscle, MA = malleus, SM = stapedial muscle, SP = sigmoid process, ST = stapes, TB = tympanic bulla, TL = tympanic ligament, TT = tensor tympani muscle.

is attached by two thin, flat pedicles (Fig. 114: AP, PP) placed respectively anteriorly and posteriorly and having their planes approximately at right angles. The cavity of the bulla is continuous with that of the middle ear and its associated air sinuses.

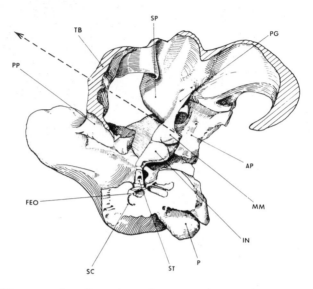

Fig. 114 Diagram of a dissection of the ossicular chain of the humpback whale, *Megaptera novaeangliae*, the dotted line represents the line of traction of the tympanic ligament.
AP = anterior pedicle, FEO = fenestra ovalis, IN = incus, MM = manubrium mallei, P = periotic, PG = processus gracilis, PP = posterior pedicle, SC = semicircular canal, SP = sigmoid process, ST = stapes, TB = tympanic bulla.

In odontocetes the anterior pedicle is absent and the posterior support does not involve bony fusion of the tympanoperiotic. A characteristic feature of the bulla of all cetaceans, recent and fossil, is a wing-like process protruding from the anterior border of the tympanic anulus, called the sigmoid process (Fig. 113: SP).

B. Tympanic membrane

In Figs 113, 115, 116 the tympanic cavity has been opened to expose the middle-ear mechanism. In both toothed and whalebone whales the meatus widens out mesially to terminate at the tympanic membrane. The eardrum

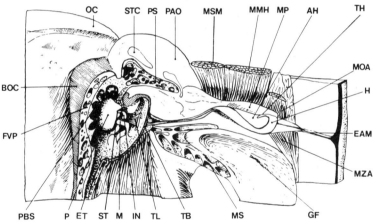

Fig. 115 Photograph and diagram of a dissection of the external meatus of *Tursiops truncatus*. The tympanic bulla has been opened to show the relationship of the meatus to the "ear drum" or tympanic ligament. The figure also shows the relationship of the middle air sinus to the glenoid fossa which prevents sound from the lower jaw reaching the middle ear.

AH = antihelix, BOC = basioccipital bone, EAM = external auditory meatus, ET = Eustachian tube, FVP = fibrovenous plexus, GF = glenoid, H = helix, IN = incus, M = malleus, MMH = mastohumeralis muscle, MOA = occipitoauricularis muscle, MP = mastoid process, MS = middle sinus, MSM = sternomastoid muscle, MZA = zygomaticoauricularis muscle, OC = occipital condyle, P = periotic, PAO = paroccipital process, PBS = peribullary sinus, PS = posterior sinus, ST = stapes, STC = styloid cartilage, TB = tympanic bulla, TH = tail of the helix, TL = tympanic ligament.

Fig. 116 Photograph of the middle ear of the Pilot-whale, *Globicephala malaena*, to show the mechanical coupling of the tympanic ligament and the tensor tympanic muscle to the malleus. Vibration of the tympanic ligament produces torsional vibration of the processus gracilis which is osseously fused to the sigmoid process. The cochlea has been opened to show the spiral labyrinth. Compare with Fig. 122 of *Platanista*.

EAM = external auditory meatus, IN = incus, M = malleus, PG = processus gracilis, S = stapes, SP = sigmoid process, TL = tympanic ligament, TT = tensor tympani muscle.

is not a thin, translucent skin, as in terrestrial mammals. It consists of two parts, a fibrous region similar to the pars flaccida of land mammals and a nonfibrous part. The fibrous portion consists of a broad, flattened, triangular "ligament" (TL) with one edge of its base attached to part of the tympanic ring, the base thus forming part of the external surface of the "membrane". The attenuated apex of the ligament, formed by the convergent radial fibres, is directed mesially and is attached to a small process on the manubrium of the malleus (M). *This ligament, hereafter referred to as the tympanic ligament, is the true homologue of the fibrous part of the tympanic membrane of land mammals.* Fleischer (1975) has stated that there is no tympanic anulus and no tympanic ligament in *Kogia breviceps*. One can only assume that this specimen was in a pathological condition for we have found both structures in *Kogia* and they are almost identical with the condition in *Physeter*. All the ziphioid whales have a similar construction. The nonfibrous part bridges the space between the fibrous part and the remainder of the tympanic ring. In *Phocoena phocoena* the nonfibrous part forms merely a few islets in the fibrous part; it is single and more extensive in area in *Globicephala melaena* and *Tursiops truncatus*. The "glove finger" (Figs 112, 117: GL) of whalebone whales is the same structure very much enlarged. Originating on the mesial aspect of the internal face of the bulla, anterior to the tympanic anulus, there is a short stout muscle, whose insertion curls laterally and merges into the fibrous matrix of the internal wall of the glove finger on the opposite side from the tympanic ligament. The homologue of this muscle in terrestrial mammals has not been found,

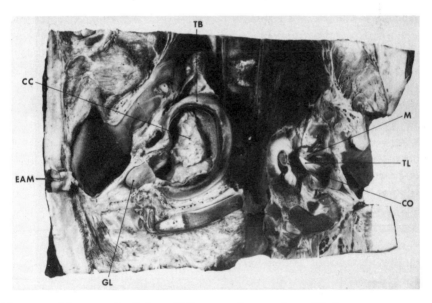

Fig. 117 Photograph of the middle ears of a mysticete (*Balaenoptera acutorostrata*) to show the corpus cavernosum tympanicum which swells under extreme hydrostatic pressure to prevent fracture of the tympanic bulla. Dissection of opposite ear shows that it is very similar in construction to that of the Odontocete. CC = corpus cavernosum tympanicum, CO = cochlea, EAM = external auditory meatus, GL = "glove finger", M = malleus, TB = tympanic bulla, TL = tympanic ligament.

but the arrangement suggests that the muscles, the glove finger, and the tympanic ligament must be regarded as a single unit, and considerably strengthens the hypothesis that the mammalian eardrum was derived from a muscle fascia of the primitive, reptilian lower jaw (Tumarkin, 1949).

C. Malleus

The three auditory ossicles found in Mammalia are also present in Cetacea. For later consideration of the mode of functioning it is necessary to describe some aspects of their structure in detail. The hammerlike shape of the malleus is maintained, but it is the processus gracilis (Figs 114, 116: PG) which resembles the handle. It will be seen that it is closely associated with the sigmoid process (SP), which forms a buttress attached to about four-fifths of its lateral border. The mesial border of the process is free from any attachment throughout its length. Between these two borders, which are thickened, is a narrowly rectangular area of much thinner bone, the borders and the thinner part forming what may be described as a channel girder. In posterior view the process forms the convexity of the girder, and it is seen that this surface is fused to the sigmoid process at a deeply grooved, arcuate junction extending from the meso-ventral to the dorso-lateral edge of the process. Examination of the junction in strong light shows that the dorso-lateral part of it is translucent, the bone in this region being extremely thin, 0·5 mm even in the adult rorqual. In general construction the odontocete processus gracilis is essentially similar to that just described for the rorqual, except that the concavity of the "girder" is not so strongly emphasized. Fusion of the processus gracilis is not peculiar to cetaceans but occurs in monkeys, carnivores, and insectivores, though in these animals there is no sigmoid process.

The manubrium of the malleus (Fig. 114: MM) in cetaceans is short and roughly conical (Mysticeti) or globular (Odontoceti). Its mass is comparable with that of the head of the malleus. At the lateral end of the inferior surface, in Mysticeti, a small promontory is identified as the processus brevis. In Odontoceti this process takes the form of a small pointed projection directed toward the head and situated about midway along the lower surface of the manubrium. In Mysticeti the attachment of the tympanic ligament extends along the whole length of the posterior face of the manubrium, whereas in Odontoceti this attachment is restricted to the processus brevis. On the proximal edge of the anterior aspect of the manubrium the point of attachment of the tensor typmani muscle is formed (Figs 113, 116: TT). The head of the malleus is deeply notched by two large facets making a re-entrant angle on its posterior aspect (Fig. 114). Both facets have smooth, shallowly convex surfaces covered with articular cartilage, which, with the corresponding facets on the incus, form part of a synovial joint. The radii of the convexities, as well as that of the arcuate junction between the two facets, lie approximately at right angles to the long axis of the tympanic ligament.

D. Incus

In general shape the incus (Fig. 114: IN) is a short, wide-based cone, whose apex is curved upward to end in the facet that articulates with a corresponding facet in the stapes (ST). The base of the cone forms the larger of two facets that articulate with the malleus. The smaller facet is approximately at right angles to the larger on the ventral aspect of the ossicle. For articulation with the malleus, both facets are shallowly concave and their line of junction is also concave. Like the facets of the malleus, these are also furnished with articular cartilage. The short process is a small conical projection directed anteriorly in line with the lateral margin of the processus gracilis of the malleus. The facet for articulation with the stapes is an oval, the long axis of which is parallel to the incudomalleolar facet.

E. Stapes

The stapes (ST) is less obviously stirrup-shaped than in most other mammals, and although an intercrural foramen exists in the rorqual, it does not occur in all cetaceans. There is no well-defined neck separating the head from the crura, and the foot is oval in shape, and is smooth, flanged, and moulded to fit precisely into the fenestra ovalis (Fig. 114: FEO). Contrary to the statements of many authors, we have been unable to find any evidence that the stapes is ankylosed to the fenestra ovalis.

F. Periotic

In cetaceans the periotic is always dissociated from the rest of the skull. During development, part of the mastoid becomes fused to the tympanoperiotic and the separated remainder to the squamosal. The amount of the mastoid which becomes fused to each of these two bones varies considerably in different kinds of cetaceans. In rorquals the much attenuated tympanic part of the mastoid process, loosely wedged between the squamosal and basioccipital bones, is maintained in position by fibrous tissue. The process of the beaked whales and some river dolphins is less attenuated but much convoluted, the convolutions interdigitating with corresponding cavities on the postero-ventral tip of the squamosal part of the mastoid. There is no fusion of the bones at this suture.

In Delphinidae the periotic part of the mastoid is neither wedged into, nor interdigitated with, that of the squamosal, and the periotic is separated from the bones of the cranium by an appreciable gap. In the region of the auditory organ, accessory air sinuses of the tympanic cavity occupy the space between the periotic and the surrounding bones of the skull, and separate the periotic from soft parts on its lateral aspect.

A study of the development of the peribullary sinuses (Fig. 111: PBS) shows a progressive investment of the periotic and invasion of the cranial bones in this area by air spaces. It is evident that these air spaces, though relatively restricted in the whalebone and beaked whales, have become very extensive in the more specialized dolphins. Thus a large hiatus (Fig. 118: H) is formed in the cranium in the region of the periotic by removal of the calcified element of bone, leaving the vascular system intact. This vascular system becomes inextricably enmeshed with the inferior and superior petrosal sinuses as well as with the cavernosum sinus, and invades the middle-ear cavity as the corpus cavernosum tympanicum (Fig. 117: CC). For this reason the intracranial part of the internal carotid artery lies entirely within the tympanic cavity, though sharply reduced in calibre, length, and function.

G. Mucous membrane

The mucous membrane of the tympanic cavity is continuous with that of the pharynx through the Eustachian tube. It invests the auditory ossicles, the corpus cavernosum tympanicum, and the muscles and nerves contained in the tympanic cavity. It forms the inner layer of the tympanic membrane in the fenestra rotunda, and lines the tympanic antrum and peribullary air sinuses. It forms several vascular folds which extend from the walls of the tympanic cavity to the auditory ossicles. One of these folds descends from the roof of the cavity to the head of the malleus and the upper margin of the body of the incus, and a second invests the stapedius muscle; others invest the chorda tympani nerve and the tensor tympani muscle. These folds separate off pouchlike recesses and give the interior of the tympanic cavity a somewhat honeycombed appearance. The membrane is pale, thin, and slightly vascular. It is covered with ciliated, columnar epithelium except over the auditory ossicles, where the cells are lower and non-ciliated. Near the anterior extremity of the cavity goblet cells are present, but otherwise there are no mucous glands. The point is deemed to have an important bearing on the function of the ossicles.

H. Middle-ear muscles

The tensor tympani muscle (TT), which is mainly tendinous, arises from the dorsal wall of the tympanic cavity near the part that transmits the Eustachian tube. It is attached in a small depression at the tip of the manubrium mallei. Although it is directed approximately in line with the long axis of the tympanic ligament (TL) as viewed ventrally (Figs 113, 116) the two attachments are displaced from each other by about 2 mm when viewed in lateral aspect. The two structures do not therefore act antagonistically, but form a mechanical couple permitting rotation of the malleus. The muscle is intimately associated with the cavernous body (CC) removed in other figures showing the middle ear. The stapedial muscle (Fig. 113: SM) is the normal position. In addition to the two muscles, we found in *Balaenoptera physalus* another short muscle occupying the groove on the anterior aspect of the processus gracilis of the malleus. It is attached by one end to the neck of the malleus and by the other to the wall of the tympanic cavity, close to the petrotympanic fissure, some of its fibres being prolonged through the fissure to reach the alisphenoid. Its nearest homologue in terrestrial mammals is the anterior ligament. In *Balaenoptera physalus* we found the muscle (Fig. 113: M) for which no homologue has yet been found.

III. Eustachian Tube and Accessory Air Sinuses of the Middle Ear.

A. Eustachian tube

The Eustachian tube (Figs 111, 118: ET) diverges from the general air cavity at a point just anterior to the tympanic bulla and passes forward

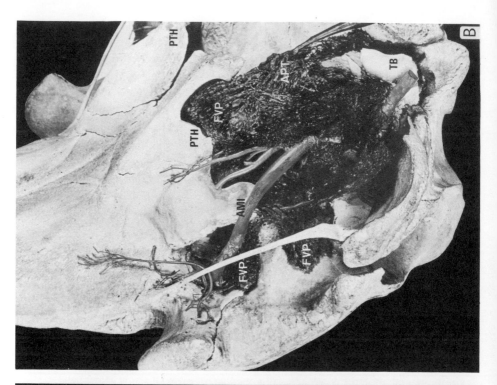

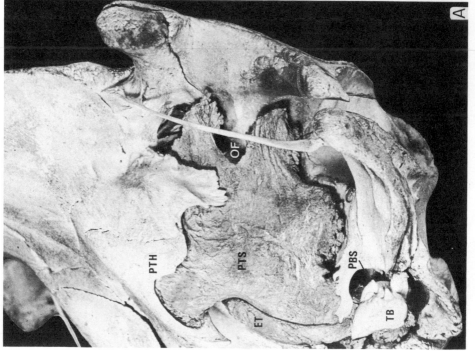

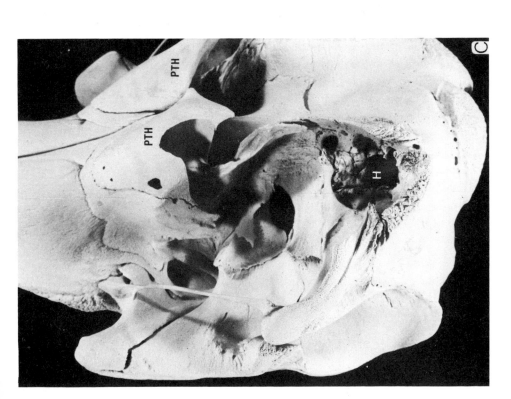

Fig. 118 (A–C) Photographs showing A: The internal polyester resin cast showing the accessory air-sac of the middle ear in *Globicephala melaena* and its relationship to the Eustachian tube. B: Polyester resin cast of the enveloping fibrovenous plexus of the air-sacs. The veins appear flattened at normal atmospheric pressure but become turgid as the air-sacs collapse under hydrostatic pressure thus filling the intervening space. C: The skull of *Globicephala melaena* showing the spaces occupied by the air-sacs and tympanic bulla.

AMI = internal maxillary artery, APT = pterygoid arterial plexus, ET = Eustachian tube, FVP = fibrovenous plexus, H = cranial hiatus, OF = orbital foramen, PBS = peribullary sinus, PTH = pterygoid hamulus, PTS = pterygoid sinus, TB = tympanic bulla.

for a short distance along the ventro-lateral margin of the basioccipital crest. Thereafter it turns mesially through a notch in this bone, and, gradually narrowing, opens into the nasopharynx. Its closing walls are made up of fibrous tissue, which frequently is partly ossified, and an internal fibrovenous plexus. The lining mucous membrane is deeply indented with valvular pockets and folds which are directed towards the choanae.

IV. Air-Sacs

The tympanic cavity and the Eustachian tube communicate with a system of pneumatic sinuses, which in volume, extent, and structure are peculiar to the cetaceans. Anterior to the tympanic bulla there is a very large air sinus which in rorquals occupies nearly the whole of the pterygoid fossa. In beaked whales and dolphins this sinus is even more extensive, covering the larger part of the base of the cranium in addition to the pterygoid fossa, which is very much enlarged. In the common dolphin the sinus has an anterior extension which passes forward on the ventral surface of the rostrum for approximately two-thirds of the latter's length. The internal cast of the air sinuses in the Pilot whale, *Globicephala melaena*, is shown and compared with same region of the skull in Fig. 118, A–C.

The deep walls of the pneumatic sinuses are closely applied to the bones of the skull. Laterally and externally they are closed by a tough, fibrous membrane resembling periosteum to which the secondarily reduced pterygoid muscles of the lower jaw are mainly attached. Within the fibrous closing membranes there is a fibrovenous plexus (Fig. 118, B: FVP) which is coextensive with the envelope of the air sinuses. The plexus is made up predominantly of large vessels which appear flattened when the sinuses are injected with air or other media, but become turgid when injected themselves under pressure. In this state they almost obliterate the cavity of the air sinuses.

A. Histology

A comparative study of the periosteal sheath and the vascular plexus shows that they are derived, respectively, from the periosteum and the Haversian system of the pterygoid bone. In the river dolphin, *Platanista gangetica*, the lateral lamina of the bone is complete; in the La Plata dolphin, *Pontoporia blainvillei*, it is present but extensively fenestrated;

and in the Amazonian dolphin, *Inia geoffrensis*, the calcified element is absent. During the evolution of cetaceans generally, it is clear that the extension of the air-sac system has involved removal of the calcified element of their contiguous bones, leaving complicated vascular systems and periosteal membranes in the peribullary, pterygoid, orbital, and maxillary regions. Lining the sacs is a thick mucous membrane which is continuous with that of the middle ear and Eustachian tube, and is remarkable for the richness of its glands and ducts. The openings of the ducts, which cover the entire inner surface of the air sinuses, are less than 0·1 mm apart and lead into a maze of smaller, racemose channels and crypts lined with columnar epithelium and goblet cells. At the entrance to the ducts, and on the exposed surface of the mucous membrane, is a layer of ciliated epithelium. Observations on freshly killed specimens reveal that the sinuses are entirely filled with an albuminous foam, formed from an oil-mucus emulsion. The oil droplets are remarkably uniform in size, with a diameter of about 1 μ.

It is clear that the main function of these air-sacs is that of equalizing the air pressure in the middle ear with that of the external hydrostatic pressure at all depths. As the air-sacs contract according to Boyle's law, the space left is taken up by the injection of blood into the fibrovenous plexus FVP through the pterygoid arterial plexus APT which is a branch of the internal maxillary artery AMI (Fig. 118).

Since oil will dissolve 16 times more nitrogen than will an equivalent amount of blood the oil droplets ensure that no nitrogen goes into solution in the vascular system. The efficiency of the air bubbles in acting as reflectors and in damping out unwanted sound vibrations is discussed in the Introduction.

The vascular system of the corpus cavernosum tympanicum (Fig. 117: CC) is supplied by the diminutive internal carotid artery and cannot be injected by normal manual pressure. However if extreme pressure is used, the corpus becomes an erectile body and swells to fill almost the entire tympanic bulla. It is conceivable that this phenomenon occurs at depth under extreme hydrostatic pressure, thus preventing fracture of the bulla. An account has been given by Fraser and Purves (1955) of fractured ear bones of Blue whales which had subsequently healed. The fracture could only have occurred through excessive hydrostatic pressure since the ear bones are remote from the surface of the body and protected by about 10 cm thickness of yellow elastic tissue. Any damage caused through harpooning would have resulted in the death of the animals. It is possible that the fracture was due to the failure of the mechanism described above.

V. Vascular System

Details of the vascular supply may be obtained from Fraser and Purves (1960), but may be summarized as follows.

The blood supply and drainage of the muscles of the external ear are those normally found in terrestrial mammals, but the meatus and its cartilages are practically nonvascular, a fact that is important in considering sound conductivity. The mucous membrane of the tympanic cavity and the corpus cavernosum tympanicum are supplied by the diminutive internal carotid artery; we agree with Reysenbach de Haan (1957) that the artery probably becomes obliterated because of the acoustic disadvantage of having a non-liberated main artery in the tympanic cavity: "The fact that its lumen has completely or nearly closed must undoubtedly be beneficial for hearing". The internal jugular vein is also reduced in size and function, and drainage of the middle ear and its structures takes place mainly through the transverse and cavernous sinuses of the cranium.

The fibrovenous plexus of the air-sacs is supplied by an arterial plexus (Fig. 118, B: APT) which emerges from the internal maxillary artery (AMI) immediately anterior to the tympanic bulla, and by small arterial branches which emerge from the same artery as it passes forward across the ventral surface of the cranium. The plexus is drained by three distinct paths: (1) by large vessels that communicate with the transverse and cavernous sinuses of the cranium and drain eventually into the spinal meningeal veins; (2) by large vessels that join the external jugular vein via the vena pterygoidea (VPT); and (3) by an intricate plexus of small veins which penetrates the fibrous covering of the sinus at the angle formed by the lateral pterygoid and tensor palati muscles. This plexus is very dense and ramifies throughout the mass of fatty tissue which lies on the mesial aspect of the lower jaw, eventually coalescing into a single vessel that joins the mandibular vein. The intramandibular fatty body is therefore not an amorphous structure, but contains an intricate network of arteries and veins which would, seemingly, interfere with the conduction of sound waves.

VI. The Ear of the Blind Indus Dolphin *Platanista indi*

A detailed description of the ear nose and throat of *Platanista indi* has been given by Purves and Pilleri (1973) so that only those features that are

relevant to the theory of cetacean hearing propounded in this book will
be described here.

A. External auditory meatus

The external auditory meatus appeared at the surface of the skin as a small,
semi-circular aperture 3 mm across the chord, with the convexity facing
postero-dorsally (Fig. 119). It was situated 26 mm posterior to the tip of
the rostrum and 16 cm above the mid-ventral line measured round the
circumference of the head. In this position it was 6 cm posterior to the
eye and 2 cm above the level of the latter, thus differing markedly in location from that of most odontocetes in which the aperture is slightly below
and posterior to the eye. The general situation of the external auditory
aperture, relatively high on the upper segment of the head, is strongly
reminiscent of the condition found in the Sirenia and indeed of that in
terrestrial mammals. The lumen of the meatus (Fig. 120) within the blubber
was relatively very wide, more so than that in any other dolphin that we
have examined, so that it was easily possible to pass a 2 mm probe through
the depth of the blubber which measured 1 cm in thickness at this point.
Below the blubber the meatus was enclosed within the mesial concavity
of a longitudinal fold of cartilage 8 mm wide (Fig. 121: AC) which curved
postero-ventrally for 6 cm along the upper margin of the zygomatic process of the squamosal bone (Fig. 120: SQ) until it reached the tympanic
bulla. The meatal cartilage was somewhat navicular in shape and is
probably representative of the antihelix of the auricle of terrestrial mammals. Apart from the gentle curvature described above, the meatal cartilage was relatively straight and not spirally twisted as it is in most
delphinids. The cartilage, moreover, was complete throughout its length
and did not show the fragmentation that can be observed in some other
delphinids. Immediately deep to the blubber, there appeared to be vestigeal
auricular muscles, namely the auricularis superior (MAS) and the zygomaticoauricularis (MZA) but these were mainly represented by large
fascia which passed imperceptibly into a mass of adipose tissue. On
cutting through the auricular cartilage longitudinally (Fig. 120: EAM)
and transversely (Fig. 121), the meatal lumen was found at its distal end
to consist of a narrow tube about 2×4 mm wide and lined by an almost
black epithelium. Towards its mesial extremity, it gradually widened to
about 4×10 mm, the epithelial lining meanwhile becoming less pig-

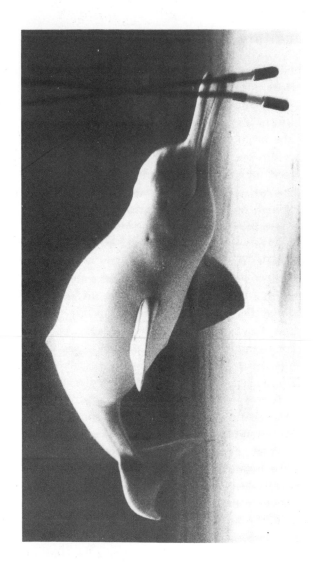

Fig. 119 Photograph of the blind *Platanista indi* inspecting hydrophones. Note the indentation at the site of the external auditory meatus. Aquarium of the Brain Anatomy Institute, Berne.

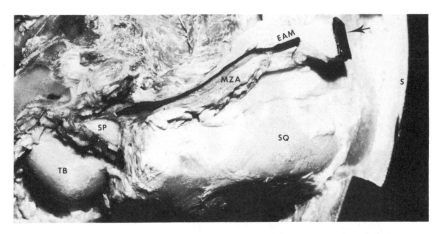

Fig. 120   Dissection of external auditory meatus leading to the tympanic bulla in *Platanista indi*. The great enlargement of the squamosal bone probably protects the meatus from damage.

EAM = external auditory meatus, MZA = zygomaticoauricularis muscle, S = skin, SP = sigmoid process, SQ = squamosal bone, TB = tympanic bulla.

mented, until it reached the tympanic bulla which was encased on its ventral and lateral aspects in a thick fibroelastic capsule. At this point it became invaginated to turn round the posterior aspect of the sigmoid process (Fig. 120: SP) where it expanded into a relatively capacious vestibule about 1 cm in diameter, and was closed mesially by the external aspect of the tympanic membrane.

B. Tympanic membrane

Buchanan (1828) described the tympanic membrane of *Monodon* as being like a convolvulus flower and this description could also be applied with certain reservations to this membrane in *Platanista indi* (Fig. 122, A). The external aspect of the membrane (TM) consisted of an almost circular hollow cone but mesially, towards its attachment to the malleus (M), the apex of the cone was flattened into a thick, triangular ligament, reminiscent of the "tympanic ligament" of most odontocetes. We consider the triangular tympanic ligament of the majority of odontocetes to be the true homologue of the tympanic membrane of terrestrial mammals, the soft distal extremity or closing wall of the vestibule being the homologue of the

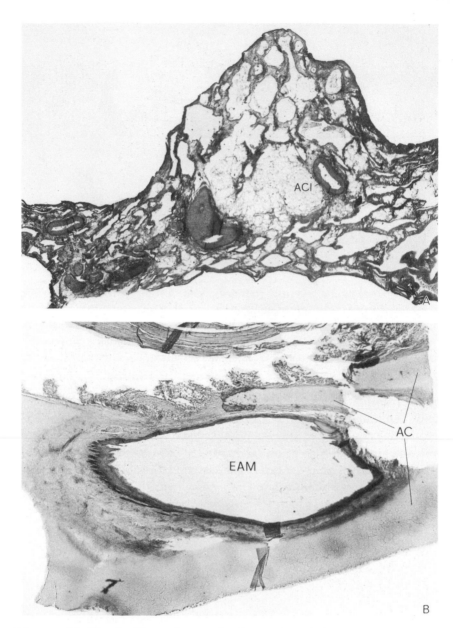

Fig. 121 (A–B) A: Photomicrograph of a section of the corpus cavernosum tympanicum of *Platanista indi* to show the large size of the veins. Probably the total role of pressure equalization in the middle ear is taken over by this organ since the accessory air-sacs are not collapsible in the species (see Fig. 122). ACI = arteria carotis interna. B: Photomicrograph of a section of the external auditory meatus of *Platanista indi* to show its large calibre compared with other odontocetes of comparable size. AC = auricular cartilage, EAM = external auditory meatus.

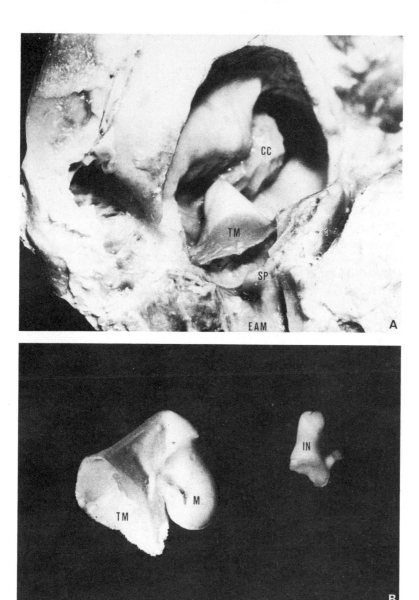

Fig. 122 (A–B) A: Photograph of a dissection of the middle ear in *Platanista indi* to show the relationship between the external auditory meatus and the tympanic membrane. The tympanic membrane is the homologue of the tympanic ligament in marine species. CC = corpus cavernosum tympanicum, EAM = external auditory meatus, SP = sigmoid process, TM = tympanic membrane. B. Isolated auditory ossicles of *Platanista indi* showing the attachment of the tympanic membrane to the malleus and the similarity of the incus to that of other cetaceans. IN = incus, M = malleus, TM = tympanic membrane.

pars flaccida. If it can be imagined that the shallow cone of the tympanic membrane of terrestrial mammals were like a parasol which could be partially closed and then laterally compressed, such would be the process of modification of this membrane in the Cetacea with the condition in *Platanista* representing a half-way stage in this process. Nevertheless, McCormick *et al.* (1970) in their experiments purporting to discount the external auditory meatus and tympanic membrane as being part of the sound conductive apparatus, stated that the "tympanic membrane" was in no way connected to the malleus and described the triangular ligament as being a suspensory ligament for the malleus.

C. Ossicles of the middle ear

The auditory ossicles in *Platanista gangetica* were described in detail by Doran (in Anderson, 1878) and this description holds almost exactly for *Platanista indi*. It can be confirmed that the apex of the tympanic membrane is attached to the recurved, mesial extremity of the manubrium of the malleus (Fig. 122, B) and that part of its distal anterior margin fits into a groove on the former's ventral aspect. The tensor tympani muscle was also, as in *Platanista gangetica* inserted into a deep pit in the dorsal aspect of the manubrium near its root. Doran's description was made from the dried ossicles after the malleus had become detached from the bulla, and he has omitted to mention its osseous fusion with that bone. This is not surprising as the fusion of the processus longus, or gracilis of the malleus to the sigmoid process of the bulla is exceedingly thin and delicate, so as to form an ideal union for high frequency vibrations. McCormick *et al.* (*loc. cit.*) on examination of the incudo-malleolar joint in a specimen of *Tursiops* found an osseous fusion between these bones although in a very limited area, and from this they deduced that the incudo-malleolar joint was a synostosis in all cetaceans. We must hasten to add that we have never found this condition of the joint in the many species that we have examined and that in the present specimens the incus (Fig. 121: IN) was found lying loose in the tympanic cavity due to the post-mortem deterioration of its mucous membrane and incudal ligament. The concavo-convex, double articulation between the malleus and incus is so constant and characteristic in shape throughout the entire Order Cetacea, notwithstanding considerable morphological differences in the shape of the bones themselves at Familial level, that it would be most strange if the bones were normally osseously fused together and the joint consequently without function.

The stapes was a small rod-like ossicle having a small posterior projection for the attachment of the stapedial muscle. Contrary to the finding of Anderson that in young specimens of *Platanista gangetica* there is an intercrural canal for the passage of the chorda tympani nerve, we could find no trace of such a canal in the present specimen notwithstanding that it was a young animal.

D. Cochlea

Anderson has given a very full description of the cochlea in *Platanista gangetica* but his figures are relatively minute and difficult to follow from the text. As it was undesirable at this stage to remove the periotic from the skull, no account will be given here of the internal auditory meatus and of the aqueducts of the cochlea and vestibule, all of which occur on the cranial aspect of the periotic. In order to ascertain whether there were any unusual features of the cochlea in *Platanista indi*, the bulbous ventral dome was removed to expose the scala vestibuli (Figs 123, 124: SCV). The fenestra ovalis was situated at the bottom of a depression on the postero-ventral margin of the cochlea and was closed by the stapes (ST) which was flanged at its base to fit the contour of the fenestra exactly. The vestibule (VE) consisted of a concavo-convex chamber the concavity of which formed the vestibular terminations of the three whorls of the semi-circular canals (SSC). Anderson states that the semi-circular canals in *Platanista gangetica* are larger in porportion to the size of the cochlea than in any other cetacean and judging by the size and spacing of their orifices in the present specimen it would seem that the size of the canals in *Platanista indi* are commensurate with those in *Platanista gangetica*.

At the mesial extremity of the vestibule was the entrance to the scala vestibuli (SCV) which wound in an elliptically spiral course for one and a half turns in a lateral and dorsal direction towards an oval ampulla where it terminated. The diameter of the scala gradually increased in calibre from $1 \cdot 5$ mm at the commencement of the ampulla downwards, being 3 mm at the termination and at the entrance to the vestibule. Notwithstanding, the basilar membrane (BM) which contained the organ of Corti was conversely decreased in width so that near the vestibule it could barely be seen with the unaided eye. At the time of the dissection no instrument

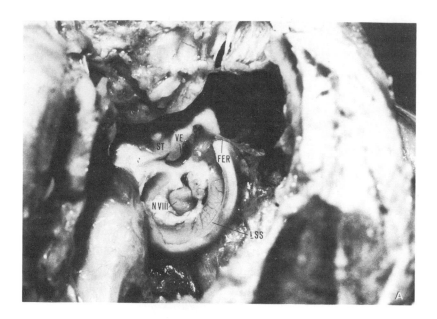

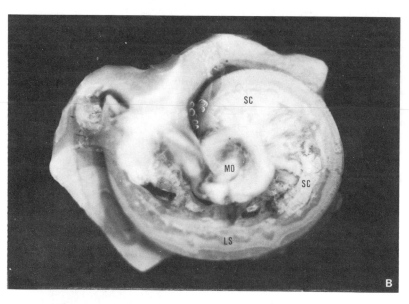

Fig. 123 (A–B)   A: Photograph of the cochlea of *Platanista indi* after removal of the base of the periotic. FER = fenestra rotunda, LSS = lamina spiralis secundaria, NVIII = acoustic nerve, ST = stapes, VE = vestibule. B: Enlarged photograph of the ventral whorl of the spiral labyrinth of *Platanista indi* to show the spiral canal which contains the spiral ganglion. LS = lamina spiralis, MO = modiolus, SC = spiral canal.

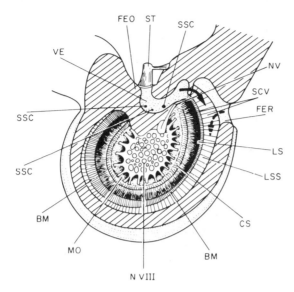

Fig. 124 Schematic drawing of the cochlea of *Platanista indi*.
BM = basilar membrane, CS = spiral canal, FEO = fenestra ovalis, FER =
fenestra rotunda, LS = internal lamina spiralis, LSS = lamina spiralis secundaria, MO = modiolus, NV = fifth nerve, NVIII = acoustic nerve, SCV =
scala vestibuli, SSC = semicircular canal, ST = stapes, VE = vestibule.

fine enough to measure the width of the membrane was available but by
extrapolation from the diameter of the scala vestibule at this point it must
have been no more than o·2 mm. The maximum diameter of the labyrinth
was 10 mm.

This extreme attenuation of the basilar membrane at the vestibular end
of the scala was due to the presence of a thin, inwardly projecting osseous
shelf on the external wall of the bony labyrinth which was named by
Hunter (1787) the lamina spiralis secundaria (LSS) and is common to all
cetaceans. The narrowness of the basilar membrane at this point is
probably an adaptation to the receipt of high-frequency vibrations. The
auditory nerve (N. VIII) gained access to the lamina spiralis (LS) by
branches which passed first through a spiral series of foramina in the
modiolus (MO). The branches then entered a narrow spiral canal (CS)
which when viewed by transmitted light appeared to be divided by transverse septa into a series of separate compartments containing ganglionic
material. Minute nervous filaments perforated the external wall of this
canal and were distributed radially to the lamina spiralis and basilar mem-

brane. Hunter (*loc. cit.*) who observed the last named spiral canal stated that it was "peculiar to the cetaceous family". At the postero-mesial margin of the auditory nerve, a small vestibular branch (NV) was distributed to the semi-circular canals but as its termination was broken off during dissection its exact distribution cannot be indicated in the diagram. In general form the cochlea of *Platanista indi* did not differ in any major respect from that of other odontocetes—although the osseous wall of the labyrinth is somewhat thinner than in most.

To complete the description of features associated with the tympanic cavity, mention must be made of the mucous membrane which enveloped all the structures hitherto mentioned including the corpus cavernosum tympanicum (Figs 121, 122: CC) which in *Platanista indi* was relatively more extensive than in other odontocetes. The corpus cavernosum tympanicum lay parallel with the tensor tympani muscle and, like that structure, was orientated towards the pharyngotympanic, anterior cone of the bulla. It invested the internal carotid artery (ACI) which was not as attenuated in *Platanista indi* as in other odontocetes—and received small capillary branches of the artery which wound throughout the tissue in a serpentine manner to form an erectile body (see Fig. 121, A). The corpus cavernosum tympanicum has the function of filling the tympanic cavity under extreme pressure conditions.

VII. The Cetacean Cochlea

A. Acoustic isolation

The cochlea has never been more explicitly depicted than by Boenninghaus (1903) long before echolocation in cetaceans was ever suspected. Figure 125 shows the cochlea of the Common porpoise *Phocoena phocoena* as seen from the cranial cavity. It is suspended in the centre of the periotic cranial hiatus (Boenninghaus, Fig. 14) by numerous chordae tendine which arise from the walls of the peripetrosal sinus. It will be seen that there is no osseous connection between the periotic and the rest of the skull, the only link with the brain being through the acoustic nerve (2). Figure 126, B is a parasagittal section through the cranium with part of the tympanic bulla removed to show the corpus cavernosum tympanicum (13). The mesial and ventral aspects of the bulla are also separated from the skull by part of the peripetrosal and the peribullary sinuses (6, 7). The whole arrange-

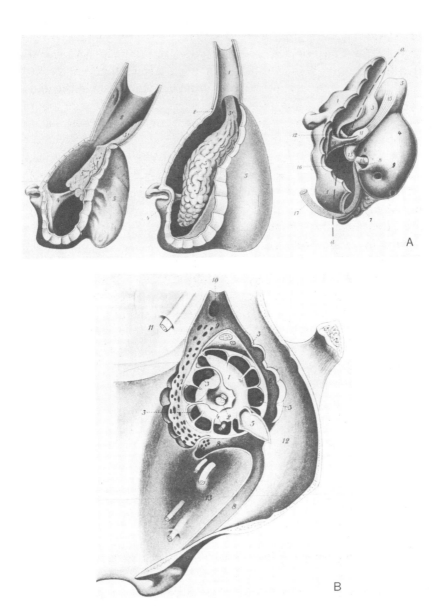

Fig. 125 (A–B) A: Drawings after Boenninghaus (1903) of dissections of the middle ear of *Phocoena phocoena* showing the corpus cavernosum tympanicum, tympanic ligament, sigmoid process, etc. For details see Boenninghaus, 1903. B: Drawing showing the suspension of the cochlea within the peripetrosal sinuses as seen from the cranial aspect.

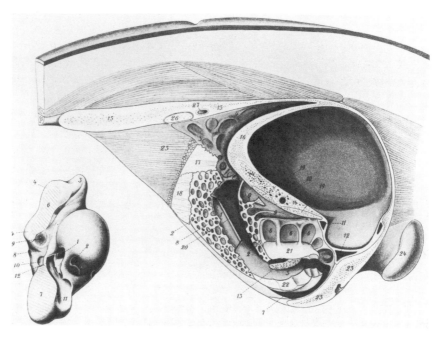

Fig. 126 Drawing after Boenninghaus of a parassagital section through the skull of *Phocoena phocoena* showing the peribullary sinuses, Eustachian tube, and fibrovenous plexus. For details see Boenninghaus, 1903.

ment is like that of a delicate microphone suspended by springs from a rigid frame. In life the various chambers of the peripetrosal and peribullary sinuses are filled with a mucoid foam containing microscopically small droplets of oil. The figures also show the fibrovenous plexus which surrounds the sinuses. It was stated that mucoid foams of this type are extremely effective in damping out sound vibrations and are morphologically highly resistant to hydrostatic pressure.

It may be inferred from this arrangement that bone conducted hearing is impossible in odontocete cetaceans since the arrangement is common to every species so far examined by us. During the course of the electro-microphonic experiments of McCormick *et al.* (1970) this delicate suspensory system was utterly destroyed and the cochlea exposed to any ambient vibration whatsoever. It is little wonder therefore that section of the external meatus, the tympanic ligament, and ablation of the malleus and mucus made "no difference" to their cochlea microphonics.

It was noted during these experiments that stimulation of one of the teeth in the direction of the long axis of the lower jaw produced a conspicuous increase in the cochlea potentials. This could be attributed to the fact that the protective, sound-damping mechanisms of the cochlea were previously destroyed. It is well known that dolphins frequently use the snout as an offensive weapon. There is a record by Gewalt (1969) of the carapace of a large turtle having been pierced in two places by dolphins using this method. The thoracic cavity of the turtle was penetrated leaving two holes 7 cm in diameter. Since the lower jaw in dolphins always protrudes beyond the snout, the total energy of the impact must have been transmitted to the area of the periotic. Imagine then, what would have happened to the cochlea had it not been protected from vibrations of the lower jaw.

The pneumatized, foam-filled, middle-sinus is present in all toothed cetaceans. It emerges from the tympanic cavity immediately anterior to the proximal end of the external auditory meatus and surrounds the mesial aspect of the glenoid fossa of the skull where the lower jaw articulates. In this position it effectively prevents the masking effect of masticatory and other mandibular noises on sound transmission in the middle ear. For this reason the lower jaw is unlikely to be the primary sound receiver as Norris postulated.

The structure of the cochlea in cetaceans is generally comparable with that of most terrestrial mammals except that it is adapted for the receipt of very high-frequency vibrations. The osseous spiral labyrinth is embedded within a relatively heavy petrosal consisting of dense petrous bone having high inertial properties, which, together with its osseous separation and suspension from the skull, reduce the likelihood of high-frequency sound vibrations reaching the cochlea by bone conduction. This acoustic isolation is still further augmented by the foam-filled peripetrosal sinuses. The whorls of the cochlea are not as tightly wound as in terrestrial mammals, there being a maximum of 2·5 in cetaceans as opposed to 3·5 in Man and even more in some other mammals. The spiral labyrinth is as usual divided into two channels by the combined osseous spiral laminae, the spiral ligament and the basilar membrane. The dorsal segment of the labyrinth, the scala vestibuli, transmits vibrations of the stapes to the cochlea fluid via the fenestra ovalis. This causes the basilar membrane to vibrate and the vibrations are further transmitted through the ventral segment of the labyrinth, the scala tympani, to the membrane of the fenestra rotunda which is contiguous with the air-filled tympanic cavity.

B. The basilar membrane

The basilar membrane incorporates the essential sound analyser — the organ of Corti. In two papers McCormick *et al.* (1970) and Wever *et al.* (*loc. cit.*) have described in detail the microscopic structure of the spiral labyrinth in *Tursiops*. The first paper is concerned with the width and length of the basilar membrane. The width varies from a minimum at the basal end of 25 μm to a maximum near the apex of 350 μm, a difference of 14-fold. This is more than twice that of the human basilar membrane which is 6.25-fold. The overall length of the membrane is comparable with that of Man, the latter being 31·5 mm and that of *Tursiops* between 35·8 and 38·5 mm.

The authors say:

> "It is of interest to compare the results obtained on the dolphin ear with those observed in Man. Retzius reported for the human ear 3475 inner hair cells and 11 500 outer hair cells for a total of 14 975 hair cells. The corresponding figures for *Tursiops* are 3451 inner and 13 933 outer hair cells for a total of 17 384 hair cells".

They report an average of 30 500 nerve ganglion cells for Man and a total of 95 004 for *Tursiops*; thus the total for the dolphin is about three times as great as that for the human ear. Of particular interest is a comparison in Man and dolphin of the ratio between nerve ganglion cells and hair cells. For the human ear this ratio is exactly 2 to 1; for the dolphin 5 to 1.

It is interesting to note that Spoendlin (1968) who worked on the cochlea of the guinea pig and cat found that only a few of the afferent, sensory neurons came from the outer hair cells, the majority coming from the inner hair cells. The outer hair cells are predominantly innervated by the efferent inhibitory neurons.

Since the outer hair cells are in the majority, one must assume that during the primary coding of the acoustic message, inhibitory processes predominate.

One of the most striking features of the dolphin basilar membrane is its extreme attenuation at the end of the basal turn of the spiral labyrinth. Here it is only 25 μm in width, less than one third the minimum of 80 μm for the human ear. Schevill and Lawrence (1953) obtained behavioural evidence of hearing in *Tursiops truncatus* for tones up to 126 kHz and

Johnson (1966) was able to determine an intensity threshold at 150 kHz. Audiograms for various species of cetacean are shown in Fig. 127.

A most extraordinary finding is that of the width of the basilar membrane in the Mysticeti. Figure 128 shows a relatively thick section through the cochlea of a 20 m Fin-whale, *Balaenoptera physalus*. The section was not thin enough to carry out a detailed examination of the organ of Corti but such sections are now available (Fig. 129).

In weight and volume the periotic is approximately eight times that of *Tursiops* and the average diameter of the osseous spiral labyrinth correspondingly large. However, we were able to measure the width of the basilar membrane at the point BM in the diagram. This was 85 μm, i.e. about the width of the basilar membrane of Man at its narrowest point. The point BM is remote from the vestibule, so that it is conceivable that

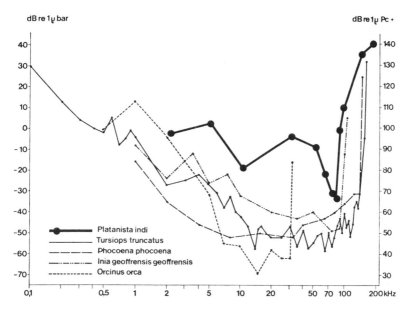

Fig. 127 Comparison of pure tone hearing threshold for five different species of dolphins.

the above figure does not represent the minimum diameter in *Balaenoptera physalus*.

If the width of the basilar membrane is correlated with the frequency range of hearing as is suggested by the observations on *Tursiops truncatus*, then the whale must be able to perceive sounds to at least the upper limit of hearing in Man. The extra diameter of the osseous spiral labyrinth is

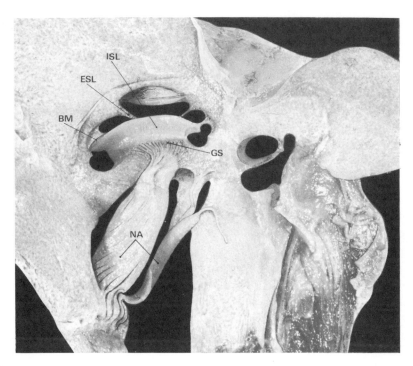

Fig. 128 Cross-section through the cochlea of the fin-whale, *Balaenoptera physalus*, to show the small diameter of the basilar membrane and the great augmentation of the osseus spiral lamina.
BM = basilar membrane, ESL = external lamina spiralis, GS = spiral ganglia, ISL = internal lamina spiralis, NA = acoustic nerve.

bridged by great augmentation of both the inner and outer spiral laminae ISL, ESL. So far no sound emissions near the upper limit of human hearing, i.e. 20 kHz, have been recorded from the Fin-whale, but Beamish and Mitchell (1971) have recorded what they call echolocating clicks from the Blue-whale, *Balaenoptera musculus*, with a dominant frequency of 25 kHz. It would be interesting to examine the larynx of the Blue-whale to try to account for these high frequencies. Fish *et al.* (1974) in their discussion of the sounds emitted by the Californian Grey-whale, *Eschrichtius robustus*, state:

"We do not know how any of the sounds discussed in this paper were actually produced by the gray whales. The metallic sounding pulsed signal produced by Gigi at Sea World sounded like bubbles escaping from an area of high pressure through a constriction. Since this sound generally was not associated

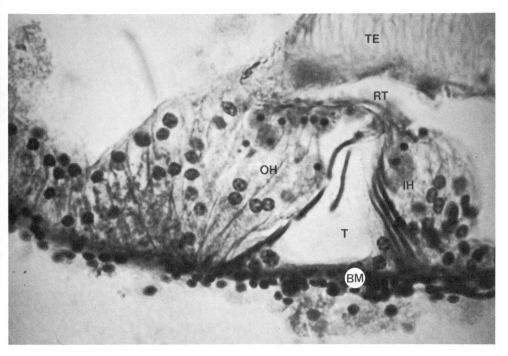

Fig. 129   Histological preparation of the organ of Corti of *Balaenoptera*
physalus.
BM = basilar membrane, IH = inner hair cells, OH = outer hair cells, RT =
reticular membrane, T = tunnel of Corti, TE = tectorial membrane.

with exhalation or blowhole movement, if it were in fact generated by escaping
air, the air must have passed from one internal chamber to another. No bubbles
were observed coming from the mouth or blowholes".

The only pressure chamber and constriction present in the respiratory
tract of the Mysticete are those of the laryngeal sac and epiglottic spout
which strongly suggests that the larynx was the sound source.

VIII. Experimental and Theoretical Considerations

The auditory meatus in cetaceans is always of small calibre relative to that
of terrestrial mammals of comparable size. There is no necessity to have a
wide lumen open to the surrounding water although this does occur in

some of the more primitive species. In the large plankton feeding whales
there is a lumen which penetrates the thick layer of blubber, but thereafter
it is completely closed as far as the ear drum. The reason for this is that the
auditory meatus is required to conduct *longitudinal* vibrations, not the
transverse waves such as are experienced by land mammals. It is only
required that the meatus should be a better conductor of sound than the
rest of the body. That such superiority obtains has been demonstrated by
Purves and Pilleri (1978). There are a number of auricular muscles
attached to the cartilage of the meatus below the blubber layer and from
their size they appear to be functional.

The dissection of *Tursiops* shown in Fig. 110 was carried out on very
fresh material, and the structures were therefore considered suitable for
sound-conductivity experiments. For this purpose an acoustic probe was
so placed that its apex touched the proximal extremity of the auditory
meatus, near the tympanic anulus, as shown in the picture (PR). This
probe was then connected to a cathode follower, an amplifier, an oscilloscope and a voltmeter. The muscles and structures in the vicinity of the
meatus were then touched at various points with another barium titanate
probe, to which was connected the output of the oscillator. Two sets of
voltmeter readings were taken at 10 and 70 kcps, respectively, at identical
points on the dissection, and it soon became obvious that at both of these
frequencies the cartilaginous meatal tube was by far the superior sound
conductor. The decibel scale of the voltmeter used a standard of 1 mW
into a 600-ohm line as zero decibels. This corresponds to 0·774 vac after
amplification of the received signal through 80 dB, and thus represents
an intensity level slightly above the threshold of human hearing.

Table 10 shows that the sound intensity received from points on or
near the meatus, especially those on the auricular muscles, are at least
10 dB above those a few cm away from the tube. Points near the insertion
of the sterno-mastoid muscle also gave good readings. It will be recalled
that the auricularis posterior muscle of terrestrial mammals originates in
close proximity to the insertion of the sternomastoid. Although no sign of
an auricularis posterior could be found in *Tursiops*, it is probable that the
fibrous sheath of the antihelix, which connects this structure with the
mastoid bone, is composed of the remnant of the fascia of this muscle.
Figure 110 shows that none of the more distant points on the external
musculature are significantly farther away from the tympanic anulus than
the distal end of the meatus itself. Many of them are nearer. This difference
of 10 dB in intensity cannot be accounted for on the assumption that all

Table 10:   Results of sound-conductivity experiments on the ear of
Tursiops truncatus

Position[a]	dB above reference at 10 kcps	dB above reference at 70 kcps
1	2·5	1·5
2	3·0	4·0
3	6·0	4·0
4	0	0
5	0	0
6	0	0
7	1·5	4·0
8	2·5	4·5
9	13·0	16·0
10	2·0	1·5
11	3·5	2·0
12	15·0	14·5
13	5·0	2·5
14	2·0	2·5
15	2·0	2·5
16	0	1·0
17	0	1·5
18	0	1·5
19	1·0	3·0
20	12·0	14·0
21	15·0	15·0
22	12·0	14·5
23	14·0	16·0
24	15·0	16·0

[a] See Fig. 110 for location of points.

the soft structures in the vicinity of the ear behave alike with respect to
the attenuation, as Reysenbach de Haan (1957) asserts. During these
experiments we found that the sound conductivity from one type of soft
structure to another, and from soft structure to bone, was relatively poor.
For instance, blubber, though an extremely good conductor, does not
readily transmit sound to the underlying muscles. We found that the
attenuation through 60 cm of blubber alone was less than through 2 cm of
blubber plus 1 cm of subjacent muscle. Table 10 shows that points

stimulated on the zygomatic process of the squamosal bone (PZS) all gave zero decibels transmission to the tympanic anulus. These differences in transmission had nothing to do with the presence of air bubbles in the tissues, but depended on the inherent variability of the molecular arrangement of the structures involved.

Johnson (1966) found that the threshold of hearing in the dolphin was about the same as that in Man except that the threshold frequency was raised by about 50 kHz. At threshold the human ear is sensitive at the ear drum to a displacement amplitude of 10^{-9} cm $= 0.1$ Ångström unit, or equivalent to 1/30 of the diameter of a molecule of oxygen.

Since the displacement amplitude for the same intensity and frequency in water is approximately 1 : 60 of that in air, then the cetacean would be sensitive to a displacement amplitude at the ear-drum of 0.016 Å if there existed no mechanism for amplifying the sound before it reached the essential organ of hearing. A specimen of the blind Indus dolphin was able to locate and touch with its snout a pellet of lead shot 3 mm in diameter (Fig. 130) suspended from a nylon thread at a distance of $1\frac{1}{2}$ m (Pilleri *et al.*, 1976). The highest echolocating frequency used during this exploration was 80 kHz i.e. with a wavelength of 1.8 cm. Consequently a great deal of the echolocating sound energy would be scattered and not returned as a direct echo. Rayleigh has shown that the *amplitude* of the *echo* varies directly with the volume of the "scatterer" and inversely as the square of the wavelength of the incident sound. That is, *the intensity* of the sound returned to the observer *varies inversely as the fourth power of the wave length*.

The ratio of the scattered to the incident amplitude must necessarily vary directly as the volume V and inversely as the distance r, and in order that the result, the ratio of amplitudes $\dfrac{a_s}{a_i}$ must have the dimensions of a pure number, we must divide by λ^2, since λ is the only linear magnitude involved i.e.

$$\frac{a_s}{a_i} \propto \frac{V}{\lambda^2 r} \text{ or } \frac{I_s}{I_i} \propto \frac{V^2}{\lambda^4 r^2}.$$

Using the values in the above experiment the proportion of the "scattered" to incident amplitude is 17×10^{-5}. Since the incident energy provides the echo the intensity must be minute.

The structure of the essential organ of hearing, the cochlea, in cetaceans is very similar to that of terrestrial mammals except that it is modified to

Fig. 130 *Platanista indi* inspecting a piece of lead shot 3 mm in thickness (arrow) suspended on the end of a thin plastic thread (photograph in the aquarium of the Brain Anatomy Institute, Berne).

process sound of very high frequency. The relative pressure amplitudes for sound waves of the same intensity and frequency in water and air is $61:1$ and the relative displacement amplitudes $1:61$.

Consequently there is a large acoustic mismatch in terrestrial mammals between the *pressure* amplitude of sound received at the ear-drum and that required by the cochlea fluid. The converse is true in cetaceans. There is no mismatch between pressure amplitude of the sound received at the external end of the ear and that transmitted to the cochlea, but there is a large mismatch in the displacement amplitude. It is universally agreed that the phenomenon of acoustic matching is carried out by the three small bones of the middle ear, the malleus, incus and stapes. These bones, although small, are relatively large compared with the minute displacement amplitudes discernible at threshold by terrestrial mammals and cetaceans and their precise mechanics have never been satisfactorily explained. That they do perform this task has been confirmed experimentally, a supreme example of which is described on p. 247 in respect of the large Fin-whale.

Figure 131 (A–D) is a schematic drawing showing the leverage of the ossicles of the middle ear in Man and three types of cetacean. The amplitude of movement of each of the three ossicles is regulated by small muscles which, for the sake of simplicity, are not shown. In Man A, the pressure at the oval window of the cochlea is determined by the ratio between the area of the ear-drum and that of the foot of the stapes, as well as by the leverage of the auditory ossicles. The ratio of the area is 30: 1 and the leverage 2: 1, giving a pressure ratio between the stapes and the drum of about 60: 1 which is approximately the same pressure amplitude ratio between a sound wave of the same intensity and frequency in water and air. In B and C, showing the arrangement in an odontocete and mysticete, the malleus and incus are of the same length but the stapes is reduced in

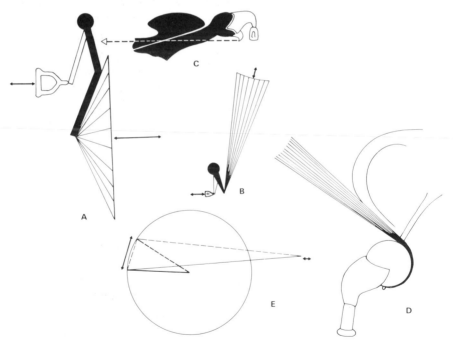

Fig. 131 (A–D)   Schematic diagrams of the auditory ossicular mechanism in a terrestrial mammal and three groups of cetaceans showing the theoretical mode of amplifying waterborne vibrations.
A: in Man, B: in the majority of odontocetes, C: in mysticete, D: in the Ziphiidae and Physeteridae.

area relative to that in Man. The ear-drum is folded like a parasol and forms a solid ligament which pulls at an acute angle to the malleus. In the Beaked-whales and Sperm-whale D the malleus is almost spherical in shape and the tympanic ligament wraps round it like a rope round a pulley. In both cases, the movements at the external end of the ligament are amplified in a cranklike manner as shown in E. The loss of pressure amplitude involved in this arrangement is compensated for by adjustments in the cross-sectional areas of the stapes and the tympanic ligament. In neither terrestrial mammals nor cetaceans is there a fixed pivot for the malleus and incus as would appear from A and B. The two bones are articulated by a synovial joint as in D, and act in a screw-driver fashion. As with the screw-driver only rotational movements are transmitted to the incus, so that in the event of a sharp shock from some other direction the bones become temporarily disarticulated, thus avoiding damage to the cochlea.

The question arises about the capacity of the relatively massive auditory ossicles of whalebone whales to undergo high-frequency movements of the extremely small amplitudes involved in the propagation of underwater sound. For the investigation of this problem a thin steel wire was soldered to the end of a varium titanate transducer, and the other end was attached to the tip of the manubrium of the malleus of a Fin-whale at the normal point of attachment of the tympanic ligament, so that it simulated the latter in length and position. For this purpose a small hole was drilled through the malleus and a steel eyelet cemented in position. The angle of attachment of the wire could be altered by raising or lowering the transducer relative to the position of the manubrium over a friction-free pulley, while the tension was kept constant by attaching a small weight to the cable connecting oscillator and transducer. The incus was allowed to rest on the malleus in its natural position, separated from the latter by a thin film of petroleum jelly. The stapes was simulated by the stylus of a microgroove crystal cartridge, which was connected to an amplifier and an oscilloscope (Fig. 132). The frequencies used in this experiment lay between 10 and 100 kHz. A considerable difference was noted in the height of the deflection of the time base in relation to the angle that the wire made with the long axis of the manubrium mallei. When the wire was pulling at a sharp angle, approximately 5 degrees, the deflection was about ten times the height attained when the wire was pulling at right angles to the manubrial axis. The only acceptable interpretation of this evidence is that the malleus was being thrown into torsional vibration, and that the manubrium was behaving like a crank actuated by the piston-like movements of the crystal

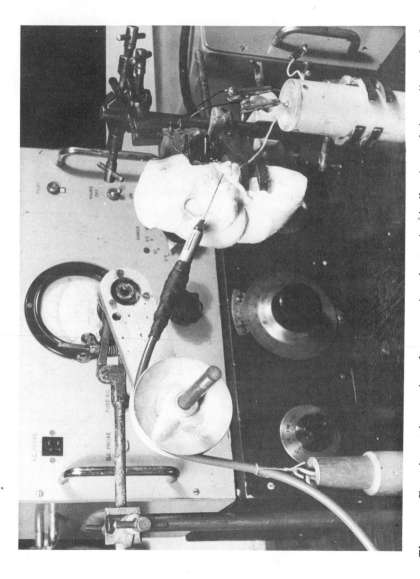

Fig. 132 Experimental procedure for measuring the torsional vibrations of the malleus of a fin-whale, *Balaenoptera physalus*. The vibration of the tympanic ligament is simulated by a barium titanate transducer and that of the stapes by the stylus of a microgroove crystal cartridge.

face. When the wire was pulling at right angles to the manubrium, the relationship between the pressure-displacement amplitudes of the crystal face and the manubrium would be approximately unity, whereas when the wire was pulling at a sharp angle to the manubrium, the pressure ratio would be reduced and the displacement ratio would be increased. Thus in the middle ear of the cetacean there exists a mechanism for increase of displacement amplitude of water-borne sounds. This method of amplification is self-compensating, as the smaller the displacement amplitude of the sound wave, the greater the relative amplification. It has been postulated by some cetologists that the middle ear of cetaceans is degenerate and that the animals hear by bone or tissue conduction. To this we may reply that anyone who has studied these bones in detail cannot avoid being struck by the sheer beauty and ingenuity of the device which has remained constant throughout the whole Order Cetacea for 15 million years whilst other structures relating to the sense of smell have disappeared completely within a much smaller space of time.

Bone-conducted hearing can be achieved in Man because the whole head is free to vibrate in an elastic medium like air, but water is virtually inelastic and incompressible.

In the cochlea the vibrations of the cochlea fluid must be greater than those of the rest of the body in order that there should be any sensation of hearing at all. We consider that there *must* be an amplification mechanism in the cetacean like that described above since in these animals hearing takes place entirely under water.

IX. Conclusion

The stories of the ancient Greeks and Romans about dolphins, regarding their affinity to Man and considered hitherto as pure mythology have now been demonstrated to lie within the bounds of actual possibility. Nowadays a new mythology has arisen and been promulgated throughout the world endowing cetaceans with attributes quite unknown amongst the rest of the Mammalia.

Notably among these are that they possess a complex communication system akin to, or even superior to, that of Man. They are thought to vocalize through the nostrils and hear through the lower jaws in the manner of snakes, although in the latter this hypothesis has been inferred from anatomical evidence and not by experimental proof. They are also said to possess an acoustic lens in the forehead for the production of directional beams of sound, the geometry and chemistry of which can be altered more or less instantaneously to change the focus of the lens at will, according to the requirements of accurate echolocation.

The recent discovery of large amounts of the glycerides of isovaleric acid in the adipose tissue of the "melon" and in the vicinity of the lower jaw has been used to support these ideas since they are not known to occur in other mammals. These deposits of fat have accordingly been named "acoustic tissues". We have suggested an alternative reason for the presence of these tissues. The present authors have adopted the simplistic view that cetaceans, like all other mammals, vocalize with the larynx and hear through the ears. It is well known from elementary acoustics that *all* high-frequency sounds, whatever their source, have directional properties provided that the wavelength of the vibrations is small compared with the linear dimensions of the source.

We suspect that the above ideas have been arrived at by the exclusive research on small cetaceans which can conveniently be kept in dolphinaria, the classic example of which is the Bottle-nosed dolphin *Tursiops truncatus*. From a comprehensive study of many species of echolocating cetaceans we find that some of them simply do not possess the alleged apparatus for producing intranarial sound nor even a well-defined "acoustic lens". With regard to hearing through the lower jaws what happens in respect to the great whales? In the Sperm-whale the so-called acoustic tissues weigh several thousand kg and yet the essential organ of hearing is no larger than that of the much smaller Killer whale. The lower jaws of the great baleen whales, the Mysticeti, can weigh in excess of 1 tonne, are constructed of solid bone throughout, and measure rather more than 20 cm in diameter. We have already demonstrated that these animals are probably capable of hearing sound frequencies equivalent to those of the upper limit of hearing in Man and probably higher.

It can be argued that the various species of cetacean vocalize and hear in different ways, but what of the construction of the larynx and ear? These are broadly similar to those of terrestrial mammals except that they have been profoundly modified for operation under water. These modifications are almost identical throughout the entire Order Cetacea and from the palaeontological evidence have remained unchanged from Tertiary times to the present day.

With regard to communication, many tables have been compiled of sounds emitted by cetaceans that are audible to the human ear, but these

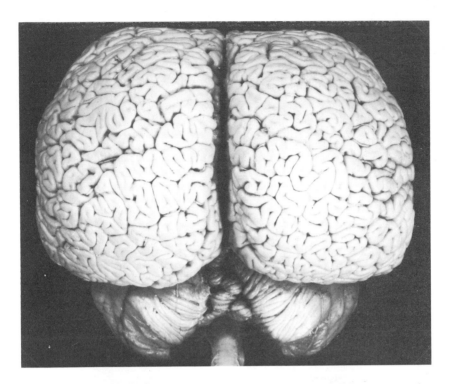

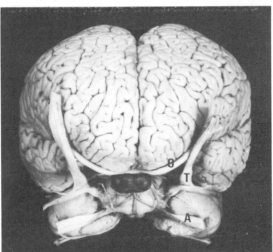

Fig. 133 Dorsal and frontal view of the brain of the bottle-nosed dolphin, *Tursiops truncatus*. O = optic, T = trigeminal, A = acoustic nerve.

could all be contained on one half of a page of a small dictionary and the repertoire is no greater than that of many marine mammals and song birds. It may be supposed that communication is derived from the echolocating pulses but we find these to be remarkably uniform in structure for any one species.

It has often been stated that dolphins "see with their ears" and this must be especially so in the blind dolphin *Platanista*. When we perceive objects with our eyes we are looking at *reflected* light. Similarly when dolphins perceive objects they are hearing *reflected* sounds from their echolocating pulses and the frequency structure and quality of such echoes must be almost infinitely great in variety. It is conceivable that a whole school of dolphins can derive information from the echoes elicited by a single member of the school and that herein lies the communication content. Most of the experimental work described in this book has been carried out with a view to establishing the *source* of cetacean vocalization feeling that most of the current ideas on this subject were erroneous. No attempt has been made to decodify a complex language.

The really spectacular and mysterious feature of the cetacean is the immensely large size and complexity of the brain which is comparable with that of human beings (Fig. 133). What proportion of the cerebral cortex is involved in the storage and processing of acoustic information is not known, but it must be quite considerable judging by the large size of the acoustic nerve. It seems a pity that so much time and energy should have been spent on the analyses of sounds emitted by dolphins rather than the study of echoes from the echolocating pulses.

References

Agarkov, G. B., Khomenko, B. G., Manger, A. P., Khadzhinsky, V. G. and Bronsky, A. A. (1979). Upper respiratory tracts. System of epicranial respiratory passages. *In:* "Functional Morphology of Cetacea" (ed. Prof. G. B. Agarkov), 222 pp. Scientific Thought, Kiev.

Anderson, J. (1878). Anatomical and Zoological Researches; comprising an Account of the Zoological Results of the two Expeditions to Western Yunnan in 1868 and 1875 and a Monograph of the two Cetacean Genera *Platanista* and *Orcaella*. 2 Vols. B. Quaritch, London.

Batteau, D. W. (1966). Theories of Sonar Systems and their Application to Biological Organisms. *In:* "Les Systèmes Sonars Animaux" (ed. R. G. Busnel), Vol. II, pp. 1033–1092.

Beamish, P. and Mitchell, E. (1971). Ultrasonic sounds recorded in the presence of a blue whale *Balaenoptera musculus*. *Deep-Sea Res.* **18**, 803–809.

Beauregard, H. (1894). Recherches sur l'appareil auditif chez les mammifères. 3 Oreille moyenne. *J. Anat. Physiol.* (*Paris*) **30**, 366–413.

Boas, J. E. V. (1912). Ohrknorpel und äusseres Ohr der Säugetiere. Eine vergleichend—anatomische Untersuchung. 226 pp. + 25 Pl. Nielsen and Lydicke, Kopenhagen.

Boenninghaus, G. (1902). Der Rachen von *Phocaena communis* Less. Eine biologische Studie. *Zool. Jahrb.* (*Anatomie*) **17**, 1–92.

Boenninghaus, G. (1903). Des Ohr des Zahnwales, zugleich ein Beitrag zur Theorie der Schalleitung. *ibid.* **19**, 189–360.

Boyle, R. W., Lehmann, J. F. and Reid, C. D. (1926). Detection of Icebergs by an Ultrasonic Beam. *Trans. Roy. Soc. Canada, 3rd Ser.*, **21**, 233–245.

Brown, D. (1962). Further observations on the pilot whale in captivity. *Zoologica* (*N.Y.*) **47**, 1–98.

Buchanan, T. (1828). "Physiological Illustrations of the Organ of Hearing." London.

Bullock, T. H. and Gurevich, V. S. (1979). Soviet Literature on the Nervous System and Psychobiology of Cetacea. *Int. Rev. Neurobiol.* **21**, 47–127.

Burns, D. (1921). "Introduction to Biophysics," pp. 325–330. J. and A. Churchill, London.

Carte, E. and MacAlister, A. (1868). On the Anatomy of *Balaenoptera rostrata*. *Phil. Trans.* **158**, 210–261.

Clarke, M. R. (1979). Der Kopf des Pottwals. Spektrum der Wissenschaft, pp. 20–28.

Clarke, M. R. (1981). Personal communication.

Clay, C. S. and Medwin, H. (1977). "Acoustical Oceanography: Principles and Applications," 544 pp., John Wiley and Sons, London.

Diercks, K. J., Trochta, R. T. and Evans, W. E. (1973). Delphinid sonar: measurements and analysis. *J. Acoust. Soc. Amer.* **54**, 200–204.

Dubrovskii, N. A., Krasnow, P. S. and Titov, A. A. (1971). On the Emission of Echo-Location Signals by the Azov Sea Harbor Porpoise. *Sov. Phys. Acoust.* **16**, 444–447.

Evans, W. E. (1973). Echolocation by marine delphinids and one species of freshwater dolphin. *J. Acoust. Soc. Am.* **54**, 191–199.

Evans, W. E. and Prescott, J. H. (1962). Observations of the sound production capabilities of the bottle-nosed porpoise. A study of whistles and clicks. *Zoologica* (*N.Y.*) **47**, 121–128.

Evans, W. E., Sutherland, W. W. and Beil, R. G. (1964). Directional characteristics of Delphinid sounds. *In:* "Marine Bio-acoustics" (ed. W. N. Tavolga), Vol. I, pp. 353–372. Pergamon Press, New York and Oxford, 1964.

Fish, J. F., Sumich, J. L. and Lingle, G. L. (1974). Sound Produced by the Gray Whale, *Eschrichtius robustus*. *Marine Fish. Rev.* **36**, 38–45.

Fleischer, G. (1975). Über das Spezialisierte Gehörorgan von *Kogia breviceps* (Odontoceti). *Z. Säugetierkunde* **40**, 89–102.

Fraser, F. C. and Purves, P. E. (1953). Fractured Earbones of Blue Whales. *The Scottish Naturalist* **65**, 154.

Fraser, F. C. and Purves, P. E. (1960). Hearing in Cetaceans. *Bull. Brit. Mus.* (*Nat. Hist.*) **7**, 4–140.

Gewalt, W. (1969). Erste Duisburger Delphinerfahrungen an *Tursiops truncatus* Mont. *Der Zool. Garten* **36**, 268–304.

Green, R. F. (1972). Observations on the anatomy of some cetaceans and pinnipeds. *In:* "Mammals of the Sea, Biology and Medicine" (S. H. Ridgway ed.), pp. 247–297.

Green, R. F., Ridgway, S. H. and Evans, W. E. (1980). Functional and descriptive anatomy of the Bottle-nosed Dolphin nasolaryngeal system with special reference to the musculature associated with sound production. *In:* "Animal Sonar Systems" (ed. R. G. Busnel), pp. 199–249. Plenum Press, New York and London.

Gurevich, V. S. and Evans, W. E. (1976). Echolocation discrimination of complex planar targets by the Beluga (*Delphinapterus leucas*). *J. Acoust. Soc. Am.* **60**, Suppl. 1, 55–56.

Hanke, H. (1914). Ein Beitrag zur Kenntnis der Anatomie des äusseren und mittleren Ohres der Bartenwale. *Jenaische Z. Naturwiss.* **51**, 487–524.

Heezen (1957). See Clarke, M. R. (1979).

Hosokawa, H. (1950). On ceteacean larynx with special remarks on the laryngeal sac of the Sei-whale and the arythenoepiglottideal tube of the Sperm whale. *Sci. Rep. Whales Res. Inst.*, *Tokyo* **3**, 23–62.

Huber, E. (1934). Anatomical notes on pinnipedia and cetacea. *In:* "Contr. to Palaeontology, Marine mammals," (ed. E. L. Packard, R. Kellog and E. Huber), IV, pp. 106–136. Carnegie Inst., Washington.

Hunter, J. (1787). Observations on the Structure and Oeconomy of Whales. *Phil. Trans.* **77**, 371–450 (*Reprint in:* "Investigations on Cetacea" (ed. G. Pilleri), Suppl. ad Vol. XII. Brain Anatomy Institute, Berne 1981).

Husson, R. (1962). Les énormes pertes d'énergie vocale de la glotte aux lèvres et leur rôle physiologique. *La Nature* **3330**, 430–437.

Husson, R. (1963). Comme celle des chauves-souris la voix humaine contient des ultrasons. *ibid.* **3338**, 257–260.

Japha, A. (1907). Über die Haut nordatlantischer Furchenwale. *Zool. Jahrb. (Anatomy)* **24**, 1–40.

Johnson, C. S. (1966). Auditory Threshold in the Bottlenosed Porpoise (*Tursiops truncatus*, Montagu). U.S. Naval Ordinance Test Station (Nots) TP 4178, pp. 1–22.

Kacprowski, J. (1977). Physical models of the larynx source. *Arch. Acoustics* **2**, 47–70.

Kellogg, W. N. (1961). Porpoises and Sonar, 177 pp. University of Chicago Press, Chicago, 1961.

Kellogg, W. N., Kohler, R. and Morris, H. N. (1953). Porpoise Sounds as Sonar Signals. *Science* **113**, 239–243.

Kerman, J. D. Jr. and Schulte, H. (1918). Memorandum upon the anatomy of the respiratory tract, foregut and thoracic viscera of *Kogia breviceps*. *Bull. Amer. Mus. Nat. Hist.* **38**, 231–268.

Kleinenberg, S. E., Yablokov, A. V., Bel'kovich, B. M. and Tarasevich, M. N. (1969). Beluga (*Delphinapterus leucas*), Investigation of the Species, 376 pp. Transl. from Russian, Israel Progr. Sci. Transl., Jerusalem.

Krüger, F. and Schmidtke, W. (1919). Theorie der Spalttöne. *Ann. Phys.* **8**, 701–714.

Lawrence, B. and Schevill, W. E. (1956). The functional anatomy of the dolphinoid nose. *Bull. Mus. Comp. Zool. (Harvard Coll.)* **114**, 103–151.

Lawrence, B. and Schevill, W. E. (1965). Gular Musculature in Delphinids. *ibid.* **133**, 5–58.

Leakey, R. E. and Lewin, R. (1977). "Origins," 262 pp. Layout, London.

Lillie, D. G. (1910). Observations on the Anatomy and General Biology of some Members of the Larger Cetacea. *Proc. Zool. Soc. London*, 769–792.

Lilly, J. C. (1962). "Man and Dolphin," 312 pp. Doubleday and Co., London.

Lilly, J. C. (1964). Sounds emitted by the Bottle-nosed dolphin. *Science* **133**, 1689–1693.

MacKay, R. G. (1964). Deep body temperature of untethered dolphin recorded by ingested radio-transmitter. *Science* **144**, 864–866.

MacKay, S. R. and Liaw, H. M. (1981). Dolphin Vocalization Mechanism. *Science* **212**, 676–678.

McCormick, J. G., Wever, E. G., Palin, J. and Ridgway, S. H. (1970). Sound Conduction in the Dolphin Ear. *J. Acoust. Soc. Am.* **48**, 1418–1428.

Møhl, B. and Andersen, S. (1973). Echolocation: high-frequency component in the click of the Harbour Porpoise (*Phocoena ph. L.*). *Acoust. Soc. Am.* **54**, 1368–1372.

Motta, G. (1959). Il volo cieco dei pipistrelli e le pieghe ari-epiglottiche quale organo produttore degli ultrasuoni. *Atti Acc. Naz. Lincei, Ser. 8*, **5**, 83–122.

Murie, T. (1870). On Risso's grampus *G. rissoanus* (Desm.). *J. Anat. Physiol.* **5**, 118–138.

Murie, T. (1873). On the Organization of the Caaing Whale, *Globicephala melaena*. *Trans. Zool. Soc. Lond.* **8**, 235–301.

Negus, V. E. (1949). The comparative anatomy and physiology of the larynx, 480 pp. Heinemann, London.

Norris, K. S. (1964). Some problems of echolocation in Cetaceans. *In:* "Marine Bio-acoustics" (ed. W. N. Tavolga), Vol. I, pp. 317–336. Pergamon Press, New York, and Oxford, 1964.

Norris, K. S. (1968). The evolution of acoustic mechanisms in Odontocete Cetaceans. *In:* "Evolution and Environment" (ed. E. T. Drake), p. 297. Yale Univ. Press, New Haven–London.

Norris, K. S. and Harvey, G. W. (1972). A Theory for the Function of the Spermaceti Organ of the Sperm Whale (*Physeter catodon* L.). NASA SP-262 pp. 397–417.

Osterhammel, K. (1941). Optische Untersuchung des Schallfeldes kolbenförmig schwingender Quarze. *Akust. Zschr.* **6**, 73–86.

Perkins, P. J. (1966). Communication sound of Finback Whales. *Norsk. Hvalfangst Tid.* **10**, 199–200.

Pilleri, G. (1970). *Platanista gangetica*, a dolphin that swims on its side. *Rev. Suisse Zool.* **77**, 305–307.

Pilleri, G. (1974). Side-Swimming, Vision and Sense of Touch in *Platanista indi* (Cetacea, Platanistidae). *Experientia* **30**, 100–104.

Pilleri, G. (1979). "Sonar Field Patterns in Cetaceans, Feeding Behaviour and the Functional Significance of the Pterygoschisis. Investigations on Cetacea" (ed. G. Pilleri), Vol. X, pp. 147–156. Brain Anatomy Institute, Berne.

Pilleri, G. (1980). "The Secrets of the Blind Dolphins." 216 pp. Sind Wildlife Management Board, Karachi. (German edition by Hallwag, Berne, 1975).

Pilleri, G. and Gihr, M. (1979). Skull, Sonar-Field and Swimming Behaviour of *Ischyrorhynchus vanbenedeni* (Ameghino, 1891) and Taxonomical Position of the Genera *Ischyrorhynchus Saurodelphis Anisodelphis* and *Pontoplanodes* (Cetacea). *In:* "Investigations on Cetacea" (ed. G. Pilleri), Vol. X, pp. 17–70. Brain Anatomy Institute, Berne.

Pilleri, G., Gihr, M., Purves, P. E., Zbinden, K. and Kraus, C. (1976a). On the Behaviour, Bioacoustics and Functional Morphology of the Indus River Dolphin (*Platanista indi* Blyth, 1859). *In:* "Investigations on Cetacea" (ed. G. Pilleri), Vol. VI, pp. 13–14. Brain Anatomy Institute, Berne.

Pilleri, G., Zbinden, K., Gihr, M. and Kraus, C. (1976b). Sonar Clicks, Directionality of the Emission Field and Echolocating Behaviour of the Indus River Dolphin (*Platanista indi* Blyth, 1859). *In:* "Investigations on Cetacea" (ed. G. Pilleri), Vol. VII, pp. 13–43. Brain Anatomy Institute, Berne.

Pilleri, G., Zbinden, K. and Kraus, C. (1979). The Sonar Field of *Inia geoffrensis*. *In:* "Investigations on Cetacea" (ed. G. Pilleri), Vol. X, pp. 107–176. Brain Anatomy Institute, Berne.

Pilleri, G., Zbinden, K. and Kraus, C. (1980). Characteristics of the Sonar System of Cetaceans with Pterygoschisis. Directional Properties of the Sonar Clicks of *Neophocaena phocaenoides* and *Phocoena phocoena* (Phocoenidae). *In:* "Investigations on Cetacea" (ed. G. Pilleri), Vol. X, pp. 157–188. Brain Anatomy Institute, Berne.

Pouchet, G. and Beauregard, H. (1889). Recherches sur le Cachalot. *Nouv. Arch. du Mus. Paris, Sér.* **3**, 1–90.

Purves, P. E. (1955). The Wax Plug in the External Auditory Meatus of the Mysticeti. *Discovery Rep.* **27**, 293–302.

Purves, P. E. (1966). Anatomical and Experimental Observations on the Cetacean Sonar System. *In:* "Les systèmes sonars animaux" (ed. R. G. Busnel), Vol. I, pp. 197–269. Lab. Physiol. Acoust. INRA, Jouy-en-Josas (France).

Purves, P. E. (1966). Anatomy and Physiology of the Outer and Middle Ear in Cetaceans. *In:* "Whales, Dolphins and Porpoises" (ed. K. S. Norris), pp. 320–380. University of California Press, Berkeley and Los Angeles.

Purves, P. E. and Pilleri, G. (1973). Observations on the Ear, Nose, Throat and Eye of *Platanista indi*. *In:* "Investigations on Cetacea" (ed. G. Pilleri), Vol. V, pp. 13–57. Brain Anatomy Institute, Berne.

Purves, P. E. and Pilleri, G. (1979). The Functional Anatomy and General Biology of *Pseudorca crassidens* (Owen) with a Review of the Hydrodynamics and Acoustics in Cetacea. *In:* "Investigations on Cetacea" (ed. G. Pilleri), Vol. IX, pp. 68–227. Brain Anatomy Institute, Berne.

Purves, P. E. and van Utrecht, V. L. (1963). The Anatomy and Function of the Ear of the Bottle-nosed Dolphin *Tursiops truncatus*. *Beaufortia* **9**, 241–256.

Rawitz, B. (1900). Die Anatomie des Kehlkopfes und der Nase von *Phocaena communis*. Cuv. *Intern. Mschr. Anat. Physiol.* **17**, 245–354.

Reysenbach de Haan, F. W. (1957). Hearing in Whales. *Acta oto-laryngologica: Suppl.* **134**, 114 pp. Stockholm.

Ridgway, S. H., Carder, D. A. and Green, R. F. (1980). Electromyographic and pressure events in the nasolaryngeal system of dolphins during sound production. *In:* "Animal Sonar System" (ed. R. G. Busnel and J. F. Fish), pp. 239–249. Plenum Press, New York and London.

Romanenko, E. V. (1976). Acoustics and Hydrodynamics of certain Marine Animals. *Sov. Phys. Acoust.* **22**, 357–358.

Schenkkan, R. J. (1972). On the Nasal Tract Complex of *Pontoporia blainvillei* Gervais and d'Orbigny, 1844 (Cetacea, Platanistidae). *In:* "Investigations on Cetacea" (ed. G. Pilleri), Vol. XI, pp. 83–90. Brain Anatomy Institute, Berne.

Schenkkan, R. J. (1973). On the Comparative Anatomy and Function of the Nasal Tract in Odontocetes (Mammalia, Cetacea). *Bijdr. Dierkunde* **43**, 127–159.

Schenkkan, E. F. and Purves, P. E. (1973). The Comparative Anatomy of the Nasal Tract and the Function of the Spermaceti Organ in the Physeteridae (Mammalia, Odontoceti). *Bijdr. Dierkunde* **43**, 93–112.

Schevill, W. E. (1964). Underwater sounds of cetaceans. *In:* "Marine Bio-Acoustics" (ed. W. N. Tavolga), pp. 307–316. Pergamon Press, Oxford.

Schevill, W. E. and Lawrence, B. (1953). Auditory Response of a Bottle-nosed Porpoise, *Tursiops truncatus*, to Frequencies above 100 Kc. *J. Exp. Zool.* **124**, 147–165.

Schevill, W. E., Watkins, W. A. and Backus, R. H. (1964). The 20-Cycle Signals and *Balaenoptera* (Fin Whales). *In:* "Marine Bio-Acoustics" (ed. W. N. Tavolga), pp. 147–152. Pergamon Press, Oxford.

Schevill, W. E., Watkins, W. A. and Ray, C. (1969). Click Structure in the Porpoise, *Phocoena phocoena*. *J. of Mammalogy* **50**, 721–728.

Scholander, P. F. (1940). Experimental Investigations of the Respiratory Function in Diving Mammals and Birds. *Hvalraedets Skr.* **22**, 1–131.

Schulte, W. H. (1916). Anatomy of a foetus of *Balenoptera borealis*. *Mem. Amer. Mus. Nat. Hist.* **1**, 389–502.

Scott, J. C. (1966). Sound Detection Threshold in Marine Mammals. *In:* "Marine Bio-Acoustics" (ed. W. N. Tavolga), pp. 247–255. Pergamon Press, Oxford.

Spoendlin, H. (1968). Ultrastructure and peripheral innervation patterns of the receptor in relation to the first coding of the acoustic message. *In:* "Hearing Mechanisms in Vertebrates" (ed. A. V. S. de Reuck and J. Knight), pp. 89–125. J. and A. Churchill, London.

Stenzel, H. (1927). Directional properties of sound sources. *Elektr. Nachritentechnik* **4,** 239–253.

Sukhovskaya, L. I. and Yablokov, A. V. (1979). Morpho-Functional Characteristics of the Larynx in Balaenopteridae. *In:* "Investigations on Cetacea" (ed. G. Pilleri), Vol. X, pp. 205–214. Brain Anatomy Institute, Berne.

Thompson, W. D'Arcy (1890). On the cetacean larynx. Studies from the Museum of Zool., Univ. Coll., Dundee.

Tumarkin, A. (1968). Evolution of the Auditory Conducting Apparatus in Terrestrial Vertebrates. *In:* "Hearing Mechanisms in Vertebrates" (ed. A. V. S. de Reuck and J. Knight), pp. 18–40. J. and A. Churchill, London.

Van der Pol, B. (1926). Relaxation oscillations. *Phil. Trans.* May 11, pp. 978–992.

Varanasi, U., Feldmann, H. R. and Malins, D. C. (1975). Molecular basis for formation of lipid sound lens in echolocating cetaceans. *Nature* **255,** 340–343.

Watkins, W. A. (1977). Acoustic behavior of sperm whales. *Oceanus* **20,** 50–58.

Watkins, W. A. and Schevill, W. E. (1975). Sperm whale (*Physeter catodon*) react to pingers. *Deep-Sea Res.* **22,** 123–129.

Wever, E. G. (1949). "Theory of Hearing." 484 pp. John Wiley, New York—Chapman and Hall, London.

Wever, E. G., McCormick, J. G., Palin, J. and Ridgway, S. (1971). The Cochlea of the Dolphin, *Tursiops truncatus*. General Morphology. *Proc. Nat. Acad. Sci. USA* **68,** 2381–2385.

Wood, A. B. (1955). "A Textbook of Sound." 610 pp. G. Bell and Sons, London.

Wood, F. G. (1954). Underwater sound production and concurrent behaviour of captive porpoises, *Tursiops truncatus* and *Stenella plagiodon*. *Bull. Mar. Sci.* **3,** 120–133.

Zbinden, K., Pilleri, G., Kraus, C. and Bernath, O. (1977). Observations on the Behaviour and the Underwater Sounds of the Plumbeous Dolphin (*Sousa plumbea* G. Cuvier, 1829) in the Indus Delta Region. *In:* "Investigations on Cetacea" (ed. G. Pilleri), Vol. VIII, pp. 259–286. Brain Anatomy Institute, Berne.

Zbinden, K., Kraus, C. and Pilleri, G. (1978). Auditory Response of *Platanista indi*. (Blyth, 1859). *In:* "Investigations on Cetacea" (ed. G. Pilleri), Vol. IX, pp. 41–63. Brain Anatomy Institute, Berne.

Zbinden, K., Pilleri, G. and Kraus, C. (1980). The Sonar Field in the White Whale, *Delphinapterus leucas* (Pallas 1976). *In:* "Investigations on Cetacea" (ed. G. Pilleri), Vol. X, pp. 124–155. Brain Anatomy Institute, Berne.

Index